Wesam Ashour Barbakh, Ying Wu, and Colin Fyfe

Non-Standard Parameter Adaptation for Exploratory Data Analysis

Studies in Computational Intelligence, Volume 249

Editor-in-Chief
Prof. Janusz Kacprzyk
Systems Research Institute
Polish Academy of Sciences
ul. Newelska 6
01-447 Warsaw
Poland
E-mail: kacprzyk@ibspan.waw.pl

Further volumes of this series can be found on our homepage: springer.com

Vol. 229. Giovanna Castellano, Lakhmi C. Jain, and Anna Maria Fanelli (Eds.)
Web Personalization in Intelligent Environments, 2009
ISBN 978-3-642-02793-2

Vol. 230. Uday K. Chakraborty (Ed.)
Computational Intelligence in Flow Shop and Job Shop Scheduling, 2009
ISBN 978-3-642-02835-9

Vol. 231. Mislav Grgic, Kresimir Delac, and Mohammed Ghanbari (Eds.)
Recent Advances in Multimedia Signal Processing and Communications, 2009
ISBN 978-3-642-02899-1

Vol. 232. Feng-Hsing Wang, Jeng-Shyang Pan, and Lakhmi C. Jain
Innovations in Digital Watermarking Techniques, 2009
ISBN 978-3-642-03186-1

Vol. 233. Takayuki Ito, Minjie Zhang, Valentin Robu, Shaheen Fatima, and Tokuro Matsuo (Eds.)
Advances in Agent-Based Complex Automated Negotiations, 2009
ISBN 978-3-642-03189-2

Vol. 234. Aruna Chakraborty and Amit Konar
Emotional Intelligence, 2009
ISBN 978-3-540-68606-4

Vol. 235. Reiner Onken and Axel Schulte
System-Ergonomic Design of Cognitive Automation, 2009
ISBN 978-3-642-03134-2

Vol. 236. Natalio Krasnogor, Belén Melián-Batista, José A. Moreno-Pérez, J. Marcos Moreno-Vega, and David Pelta (Eds.)
Nature Inspired Cooperative Strategies for Optimization (NICSO 2008), 2009
ISBN 978-3-642-03210-3

Vol. 237. George A. Papadopoulos and Costin Badica (Eds.)
Intelligent Distributed Computing III, 2009
ISBN 978-3-642-03213-4

Vol. 238. Li Niu, Jie Lu, and Guangquan Zhang
Cognition-Driven Decision Support for Business Intelligence, 2009
ISBN 978-3-642-03207-3

Vol. 239. Zong Woo Geem (Ed.)
Harmony Search Algorithms for Structural Design Optimization, 2009
ISBN 978-3-642-03449-7

Vol. 240. Dimitri Plemenos and Georgios Miaoulis (Eds.)
Intelligent Computer Graphics 2009, 2009
ISBN 978-3-642-03451-0

Vol. 241. János Fodor and Janusz Kacprzyk (Eds.)
Aspects of Soft Computing, Intelligent Robotics and Control, 2009
ISBN 978-3-642-03632-3

Vol. 242. Carlos A. Coello Coello, Satchidananda Dehuri, and Susmita Ghosh (Eds.)
Swarm Intelligence for Multi-objective Problems in Data Mining, 2009
ISBN 978-3-642-03624-8

Vol. 243. Imre J. Rudas, János Fodor, and Janusz Kacprzyk (Eds.)
Towards Intelligent Engineering and Information Technology, 2009
ISBN 978-3-642-03736-8

Vol. 244. Ngoc Thanh Nguyen, Rados law Piotr Katarzyniak, and Adam Janiak (Eds.)
New Challenges in Computational Collective Intelligence, 2009
ISBN 978-3-642-03957-7

Vol. 245. Oleg Okun and Giorgio Valentini (Eds.)
Applications of Supervised and Unsupervised Ensemble Methods, 2009
ISBN 978-3-642-03998-0

Vol. 246. Thanasis Daradoumis, Santi Caballé, Joan Manuel Marquès, and Fatos Xhafa (Eds.)
Intelligent Collaborative e-Learning Systems and Applications, 2009
ISBN 978-3-642-04000-9

Vol. 247. Monica Bianchini, Marco Maggini, Franco Scarselli, and Lakhmi C. Jain (Eds.)
Innovations in Neural Information Paradigms and Applications, 2009
ISBN 978-3-642-04002-3

Vol. 248. Chee Peng Lim, Lakhmi C. Jain, and Satchidananda Dehuri (Eds.)
Innovations in Swarm Intelligence, 2009
ISBN 978-3-642-04224-9

Vol. 249. Wesam Ashour Barbakh, Ying Wu, and Colin Fyfe
Non-Standard Parameter Adaptation for Exploratory Data Analysis, 2009
ISBN 978-3-642-04004-7

Wesam Ashour Barbakh, Ying Wu, and Colin Fyfe

Non-Standard Parameter Adaptation for Exploratory Data Analysis

Springer

Dr. Wesam Ashour Barbakh
Computer Engineering Department
Faculty of Engineering
Islamic University of Gaza
P.O. BOX 108
Gaza, Gaza Strip
Palestine

Dr. Ying Wu
Coastal and Marine Resources Centre
University College Cork
Irish Naval Base
Haulbowline
Cobh
Co. Cork
Ireland
Email: y.wu@ucc.ie

Dr. Colin Fyfe
School of Computing
Applied Computational Intelligence
Research Unit
University of the
West of Scotland
Paisley Campus
Paisley, PA1 2BE
Scotland
UK
E-mail: colin.fyfe@paisley.ac.uk

ISBN 978-3-642-26055-1 e-ISBN 978-3-642-04005-4

DOI 10.1007/978-3-642-04005-4

Studies in Computational Intelligence ISSN 1860-949X

Softcover reprint of the hardcover 1st edition 2009

Typeset & Cover Design: Scientific Publishing Services Pvt. Ltd., Chennai, India.

Printed in acid-free paper

9 8 7 6 5 4 3 2 1

springer.com

Preface

This book presents the fruits of several years of research into non-standard methods of adaptation for exploratory data analysis. This research resulted in the award of PhD to two of the authors, Dr Wesam Barbakh and Dr Ying Wu, at the University of the West of Scotland in early 2008. Their respective areas of research are

Dr Barbakh is mainly interested in clustering and topology preserving mappings. In particular his work concentrates on formulating diverse performance functions which incorporate local and global information in order to ensure that prototypes attain a more globally optimal convergence. These non-standard performance functions have translated into non-standard adaptation algorithms.

Dr Wu is the most directly intereted in non-standard adaptation *per se*. He has extensive experience of non-standard exploratory data analysis using such methods as reinforcement learning and the cross entropy method. While most of the work reported here consists of applying such methods to linear projections, this is merely to prove that such methods are successful. These methods are applicable to any problem involving optimization.

There is increasing interest in novel methods of parameter adaptation in machine learning; this interest is mainly drawn from an interest to create true artificial intelligence. We have now many years of experience in adapting parameters using such methods as gradient descent, Expectation Maximization, Hebbian learning, competitive learning and so on. Yet while it is commonly agreed that some form of adaptation is necessary for intelligence, none of these methods – even those based on human-like adaptation – have given us software which is truly intelligent.

We begin our discussion with non-standard clustering criteria: standard clustering criteria tend to minimise a global objective or performance function whereas we mix local and global criteria in order to optimise clusterings. Our non-standard adaptation methods include

Reinforcement learning: Reinforcement learning uses trial and error to optimise a performance function and has been widely used in robotics and has even been used for problems which one tends to associate with supervised learning but there has been very little work using reinforcement learning with problems on which we usually use unsupervised learning.

Cross entropy methods: These began as a method to calculate the probability of some rare events i.e. to calculate the probability mass in the tail of some distribution but have more recently been applied to combinatorial optimisation problems. We now apply the method to exploratory data analysis.

Artificial immune systems: As with other biologically inspired algorithms such as artificial neural networks and evolutionary algorithms, these are designed to emulate the adaptation property of an existing biological system. The immune system responds to attacks from antigens and builds up for its host, an immunity towards those antigens which it has successfully defeated in the past. Thus the organism has adapted and is now more capable of future survival in its environment. We use the basic concepts from this paradigm in order to create algorithms capable of exploring large scale data sets.

We cannot claim yet that the methods of this book create a true artificial intelligence but we hope that they contribute to the discussion in this area. The individual theses of Dr. Barbakh and Dr. Wu may be found at http://cis.paisley.ac.uk/research; these certainly have contributed to the discussion on intelligence in this area but we are sure that there are still more PhDs to be had in this exciting and engrossing field and look forward to continuing collaboration with new workers in this field.

March 2008
Paisley Campus
The University of the West of Scotland

Wesam Barbakh
Ying Wu
Colin Fyfe

Contents

1

Introduction

1.1 Unsupervised Exploratory Data Analysis

It is relatively easy to give answers to well-specified questions about the statistical nature of a well-understood data set; for example, it is easy to compute probabilities and quantiles for the distribution of annual rainfall, assumed to be normally distributed, in a region. The more data there is available, the more accurate an answer can be given. However, it is extremely difficult to investigate the structure of a data set that is high dimensional and not well-understood, and in which the structures are hidden among large amounts of multivariate data. Sometimes, it seems that the more data there is available, the more difficult it is to understand the data set. Thus, to explore such a data set for new insights, methods that can discover the structure in the data set must be used and such a data-driven search for statistical insights and models is traditionally called *exploratory data analysis* (see [115, 124, 148, 222]) in the statistics literature.

We can have different goals in exploratory data analysis such as to illustrate the structure in high-dimensional data sets, or multivariate relations between data sets. Three of most common goals are visualization, clustering and projection. Visualization [135, 221] is the graphical presentation of information, with the goal of providing the viewer with a qualitative understanding of the information content, where information may be data, processes, relations, or concepts. Clustering [220, 84, 78, 28] aims to reduce the amount of data by categorizing or grouping similar data items together. One of the motivations for using clustering algorithms is to provide automated tools to help in constructing categories or taxonomies. Projection attempts to reduce the dimensionality of the data items, the goal of which is to represent the input data items in a lower-dimensional space in such a way that certain properties of the structure of the data set are preserved as faithfully as possible. A collection of projection methods [117, 86, 42, 122, 110] have been developed and applied to perform projection with different objectives.

W.A. Barbakh et al.: Non-Stand. Param. Adapt. for Explor. Data Analys., SCI 249, pp. 1–6.
springerlink.com

Generally speaking, the goal of exploratory data analysis, data and knowledge mining is learning from data and this learning from data comes in two flavors: supervised learning [38] and unsupervised learning [111]. In supervised learning, the algorithm is provided with both the input data items and the desired output. The target of the learning is to specify a relationship between the input data and the desired output as is done in regression analysis. To apply this to exploratory data analysis methods, the values of the desired output must be known for a sufficiently large part of the data set. However, there is the possibility of overfitting by memorizing all the labels for each case, rather than learning general regression relationships between the data and labels. In order to avoid over fitting, learning algorithms try to achieve a balance between fitting the training data and good generalization. In unsupervised learning, there is no desired output for the data points and all variables of the data set are treated in the same way. The task is to find a suitable representation of the underlying structure of the data. Therefore we call the process of data exploration with unsupervised learning, *unsupervised data exploration*, which enables the identification and visualisation of latent structure within an unlabeled and high-dimensional data set.

One methodology for data exploration begins by designing an objective function that can model the problem's objectives while incorporating any constraints. Gradient-based approaches [101] are most frequently used to search in the space of parameters so that the parameters can be optimized in the direction of the greatest rate of increasing the objective function value. We consider that such methods are standard adaptation methods in that the way the parameter is estimated is directly and explicitly defined according to the objective model and the adaptive update rule for the parameter is derived by using spatial and temporal partial derivatives. However, a limitation of the standard adaptation methods is that too much human effort is required in the derivation of the adaptive rules.

Non-standard adaptation is distinguishing from standard adaptation in that the learner tries to explore the global optimum of the parameters without explicitly defining an adaptive update rule. A good example of non-standard adaptation is reinforcement learning. Reinforcement learning learns what to do so as to maximize a numerical reward signal. The learner is not told which actions to take, but must discover which actions yield the most reward by trying them. Then a reinforcement learning agent exploits what it has tried in the past and found to be effective in producing reward, but it may also explore actions that it has not selected before in order to make better action selections in the future. Non-standard adaptation methods such as reinforcement learning, the cross entropy method and artificial immune systems share a popular property that they can converge to the global optimum in the seach space fast. In this thesis, we will investigate these non-standard adaptation methods in unsupervised data exploratoration, particularly, projection problems. We next discuss the overall structure of the thesis.

1.2 Projection Methods

Projection methods include such well known methods as Principal Component Analysis for a single data set and Canonical Correlation Analysis for a pair of data sets. Such methods have previously been used to compress a data set but now are more commonly used for visualization. Thus we find a linear or non-linear manifold and project the data onto this manifold in a way that reveals structure in the data set. However when used for visualization, we are limited to two or at most three dimensional representations of a data set and so the search is on for the most fruitful manifold onto which the data can be projected.

The earliest methods tended to concentrate on second order statistics and so preservation of variance was the criterion which was most used. This is a very useful heuristic as large variance often is caused by inherent structure in a data set. However it is an heuristic and so it is not guaranteed to work. Thus more recent algorithms have tended to use different criteria. However there seems to be something more needed which is the development of non-standard means of adapting the parameters. This is the central topic of this book.

1.3 Clustering

Clustering is an active area in data mining and machine learning research. It can be considered as one of the most famous and important unsupervised learning problems. Clustering is a partitioning of a data set or objects into meaningful groups or clusters. The cluster can be defined as a collection of objects which resemble each other and are dissimilar or different to the objects in the other clusters. Clustering algorithms can be used in many fields such as image processing, medicine, libraries, marketing, biology, etc. Although a large variety of clustering algorithms have been developed in the research over the last 5 decades or so, there is no single algorithm that is optimal for all different data sets. There is no unique clustering algorithm that can meet and handle all the clustering requirements. There are some problems with clustering [106]:

1. Current clustering algorithms do not address all the requirements, such as scalability, usability, ability to deal with noise and outliers, high dimensionality and insensitivity to the initial conditions, adequately.
2. Time complexity based on the number of data items and the number of dimensions.
3. The definition of "distance" affects the result of clustering.
4. If an obvious distance measure doesn't exist we must "define" it, which is not always easy, especially in multi-dimensional spaces.

There are five main categories of clustering algorithms [1, 239]:

1. Partitional algorithms.
2. Hierarchical algorithms.
3. Density based algorithms.
4. Grid based algorithms.
5. Model based algorithms.

In this thesis we consider only hierarchical and partitional clustering algorithms. More information about the clustering techniques can be found in [29, 108]. Hierarchical algorithms [125, 127] find successive clusters using previously established clusters, whereas partitional algorithms determine all clusters at once. Partitional clustering like K-means can be further subdivided into relocation algorithms and density based algorithms. Relocation algorithms try to discover clusters by relocating the prototypes in successive iterations. Density based algorithms search for areas with a high population of data.

K-means (discussed in detail in Chapter 2) is the standard method in the literature for clustering: it is extremely robust and a very simple algorithm to implement and so tends to be one of the first algorithms a data analyst will use on a new data set. It does, however, have some drawbacks, chief of which is its tendancy to find local rather than global optima depending on the initial values given to its prototypes. Thus in this thesis we search for algorithms which overcome this drawback. We tend to illustrate our methods on artificial data since we can readily see whether the clusters found are correct. With real high-dimensional data sets, we are not in a position to readily identify whether the correct clusters have been found and thus have to rely on one of the clustering indices in the literature. The fact that there are a number of clustering indices shows that there is no single unquestionable criterion which is able to robustly identify what constitutes a good cluster and what does not.

There are other problems not specifically addressed in this thesis, chief of which is determining the correct value of K i.e. how many clusters are in the data set. The data-hungry aspects of the algorithms we develop mean that any additional prototype will simply move to that part of the data set which exhibits most available data density with a subsequent rearrangment of the other prototypes. Similarly we do not address computational complexity in the thesis since all algorithms run very easily on modern PCs.

1.4 Structure of the Book

We begin the book with a review of techniques and methods related to our clustering algorithms. We begin with some of the major clustering algorithms in hierarchical and partitional categories such as agglomerative and divisive hierarchical clustering algorithms, K-means algorithm [158], K-Harmonic means algorithm, kernel K-means, etc. After that we review some famous and recent topology preserving mapping algorithms such as the self-organizing map (SOM) [137], the generative topographic map (GTM) [34], and the harmonic topographic mapping (HaToM) [184, 185].

In Chapter 3, we provide a brief survey of the most popular methods for linear and non-linear projections so that we have a general overview of different projection problems, to which the derived methods will be applied. We will also very briefly introduce one of the most popular deflation methods, Gram-Schmidt orthogonalization, that has been widely used to identify multiple components.

In Chapter 4, we create new clustering algorithms which incorporate local and global information and then self-organise based on this information. We show with extensive results that these new algorithms perform much better than e.g. K-means particularly when the initialisation of the prototypes in K-means is not ideal.

In Chapter 5, we show how to extend the clustering algorithms in Chapter 3 to make topology preserving mappings for visualization. Also we extend these algorithms and construct some new kernel clustering algorithms that work in kernel space.

In Chapter 6, we introduce new online clustering algorithms, using the same performance functions used in Chapter 4. Also we show how to use the reinforcement learning technique in clustering by introducing new reward functions. We use the same principles as in Chapter 5 to extend the new algorithms in this chapter for visualization.

In Chapter 7, we introduce a new algorithm to build the connectivity graphs. We compare this new algorithm with some famous algorithms used to build connectivity graphs. Also we show in this chapter how to use similarity functions for clustering while taking the benefits of the algorithms derived in Chapter 4.

In Chapter 8, we develop three general reinforcement learning methods for projection problems under the framework of reinforcement learning. We first derive a particular form of immediate reward reinforcement learning [231] to solve linear and non-linear projections, where the component weight vectors are represented by stochastic units drawn from the Gaussian distribution and we take adaptive update rules for the parameters of the Gaussian distribution so as to maximize the expected value of the long-term reward. Then we investigate an alternative immediate reward reinforcement learning, in which the individual unit is deterministic but has stochastic synapses. We demonstrate that the four learning rules described in Chapter 3 can be used. Lastly, we investigate two algorithms based on different temporal difference learning algorithms: Sarsa-learning [210] and Q-learning [225]. We show that the convergence is fast with high accuracy. Additionally, we describe a general method to identify multiple components by re-defining the reward functions. Two different ways are derived to identify orthogonal directions in linear projection problems and kernel methods respectively.

In Chapter 9, we start with independent component analysis to illustrate a batch cross entropy method [56]. Then we derive an on-line cross entropy method for linear projections, in which the parameters of the stochastic units are updated more smoothly. We show that with a variety of appropriate performance functions for projection problems, all the results are accurate and stable. We will also show that the method to identify multiple components described in Chapter 8 can be performed in the on-line cross entropy method, while the Gram-schmidt method is more suitable for the batch cross entropy. We consider projections based upon probability models, where we extend the maximum likelihood method and the EM algorithm in latent variable models based on cross entropy methods. We finally investigate the deep architecture in unsupervised data exploration, for which we develop a multilayer topology preserving mapping model for clustering.

In Chapter 10, we extend the artificial immune system [61] to solve projection problems. We first demonstrate that a modified CLONALG algorithm can be directly applied to independent component analysis. We show that the accuracy can be improved by setting a smaller variance when cloning and maturing the antibodies. Then we extend the modified CLONALG algorithm so that the multiple components can be identified directly by the immune system itself. We show that the non-standard adaptation algorithms we have derived or extended can be combined. We incorporate cross entropy into the immune system and integrate the immune-inspired algorithm with the Q-learning method. We compare the results from the combined algorithms with that from individual algorithms to show that the perfomance is improved not only with higher accuracy, but also with decreased size of data set and number of iterations. We also develop a boostrapping method which allow us to combine the three non-standard adaptation methods. We present a new way to determine the final solution and then show that our method converges to the optimal solution more reliably and stably.

2

Review of Clustering Algorithms

Clustering can be considered the most important unsupervised learning problem; as with every other problem of this kind, it deals with finding structure in a collection of unlabeled data. A cluster is therefore a collection of objects which are "similar" to one another and are "dissimilar" to the objects belonging to other clusters.

The K-means algorithm [109, 152, 158] is one of the most popular algorithms that can be used for clustering. It is a centre or prototype based algorithm. The main idea is to define K prototypes, one for each cluster and then distribute these prototypes in such a way as to give good clustering. It is an iterative algorithm in which K means are spread throughout the data and the data samples are allocated to the mean which is closest (often in Euclidean norm) to the sample. Then the K means are repositioned as the average of data points allocated to each mean. This continues until stable convergence is reached. The K-means algorithm is one of the first which a data analyst will use to investigate a new data set because it is algorithmically simple, relatively robust and gives 'good enough' answers over a wide variety of data sets: it will often not be the single best algorithm on any individual data set but it may be close to the optimal over a wide range of data sets [166]. However the algorithm is known to suffer from the defect that the means or prototypes found depend on the initial values given to them at the start of the simulation: a typical program will converge to a local optimum, see section (2.3.1) for more details. There are a number of heuristics in the literature which attempt to address this issue [108] but, at heart, the fault lies in the performance function on which K-means is based.

[151] proposed a global K-means algorithm, an incremental approach to clustering that adds one cluster prototype at a time through a deterministic global search consisting of N (the data size) executions of the K-means; this algorithm can obtain equivalent or better results than the standard K-means, but it suffers from high computation cost and at the same time gives no guarantee to find the optimum.

W.A. Barbakh et al.: Non-Stand. Param. Adapt. for Explor. Data Analys., SCI 249, pp. 7–28.
springerlink.com

Another possible objective of unsupervised learning can be described as topology learning or creating a topology-preserving mapping [34, 93, 161, 181]. This is a class of dimensionality reduction techniques that seek to preserve some of the structure of the data. It is a transformation which captures some structure in the data so that points which are mapped close to one another share some common feature while points which are mapped far from one another do not share this feature. The Kohonen's Self-Organizing Map (SOM) [137, 138] is a Neural Network map called a topology-preserving map because there is a topological structure mirroring that of the data imposed on the nodes in the network. It takes into consideration the physical arrangement of the nodes. Nodes that are "close" together are going to interact differently from nodes that are "far" apart. The Generative Topographic Mapping (GTM) [34, 35] was developed by Bishop as a probabilistic version of the SOM, in order to overcome some of the problems of this map, especially the lack of objective function. In the rest of this chapter, we explain some clustering algorithms like K-means and K-Harmonic means [244], and then we discuss some topology-preserving mapping algorithms like the SOM, GTM and Harmonic Topographic Map (HaToM) [182, 184].

2.1 Distance Measures

The distances between samples (or objects) are used to measure the similarities between them. All clustering methods are based on measuring the similarities between samples, so that all samples within each cluster are more closely related to one another than samples assigned to different clusters.

1. The Euclidean distance:
   The Euclidean distance is commonly used to measure the distance between two points. The Euclidean distance between $\mathbf{x} = (x_1, x_2, ..., x_D)$ and $\mathbf{y} = (y_1, y_2, ..., y_D)$ is given by:

$$d_E(\mathbf{x}, \mathbf{y}) = \| \mathbf{x} - \mathbf{y} \| = \sqrt{\sum_{i=1}^{D} (x_i - y_i)^2} \tag{2.1}$$

2. The Manhattan distance:
   The Manhattan distance between two points $\mathbf{x} = (x_1, x_2, ..., x_D)$ and $\mathbf{y} = (y_1, y_2, ..., y_D)$ is given by:

$$d_M(\mathbf{x}, \mathbf{y}) = \sum_{i=1}^{D} |x_i - y_i| \tag{2.2}$$

3. The Minkowski distance:
   The Minkowski distance between two points $\mathbf{x} = (x_1, x_2, ..., x_D)$ and $\mathbf{y} = (y_1, y_2, ..., y_D)$ is given by:

$$d_{Mk}(\mathbf{x},\mathbf{y}) = \left(\sum_{i=1}^{D}(|x_i - y_i|^P)\right)^{\frac{1}{P}} \tag{2.3}$$

where P is positive value greater than 0, $P > 0$. So d_M and d_E are special cases of d_{Mk}.

4. The Vector angle distance:
The vector angle distance between two points $\mathbf{x} = (x_1, x_2, ..., x_D)$ and $\mathbf{y} = (y_1, y_2, ..., y_D)$ is given by:

$$d_V(\mathbf{x},\mathbf{y}) = 1 - \frac{\sum_{i=1}^{D} x_i y_i}{\sqrt{\sum_{i=1}^{D} x_i^2}\sqrt{\sum_{i=1}^{D} y_i^2}} \tag{2.4}$$

5. The Correlation distance (or Pearson distance):
The correlation distance between two points $\mathbf{x} = (x_1, x_2, ..., x_D)$ and $\mathbf{y} = (y_1, y_2, ..., y_D)$ is given by:

$$d_C(\mathbf{x},\mathbf{y}) = 1 - \frac{\sum_{i=1}^{D}(x_i - \bar{x})(y_i - \bar{y})}{\sqrt{\sum_{i=1}^{D}(x_i - \bar{x})^2}\sqrt{\sum_{i=1}^{D}(y_i - \bar{y})^2}} \tag{2.5}$$

We note that every data set is isotropic in the metric with which it is clustered i.e. if we use the Manhattan distance to cluster a space, the data will be equal variance in every direction in the Manhattan metric space. However we are often in a position where there is a natural metric in a data space. For example, the Euclidean metric is optimal for a Gaussian distributed data set and many high dimensional data sets are close to a Gaussian distribution. Thus if we use a Manhattan metric in a typical data set we are de-emphasising the effect of outliers and large deviations from some central measure while if we use a Minkowski metric with $p > 2$, we are going to give more weight to outliers and data points far from the centre. Thus the choice of which metric to use in any situation will depend on the data set and the operation which we wish to apply to the data set.

2.2 Hierarchical Clustering

Hierarchical clustering algorithms [214] construct a hierarchy of nested clusters. They proceed by either a series of successive divisions or a series of successive mergers. There are two types of hierarchical clustering algorithms. The first one is called agglomerative, or bottom up, in which clusters are merged iteratively. The second is called divisive, or top down, which does the reverse of agglomerative by starting with all samples, or objects, and subdividing them into smaller groups iteratively as shown in Figure 2.1. Consider we have a data set consisting of N samples, or objects. Then the basic process of agglomerative hierarchical clustering is as follows [127]:

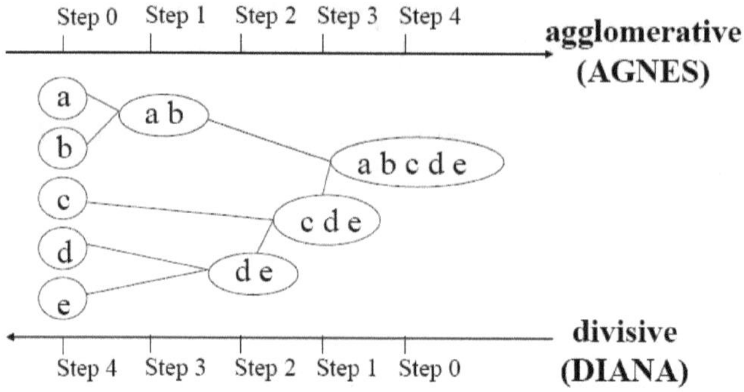

Fig. 2.1. Types of hierarchical clustering [108]

1. Compute the proximity (dissimilarity) matrix. The proximity matrix is a matrix NxN that defines the distances (or similarities) between all data points.
2. Let each data point represent a cluster.
3. Find the closest (or most similar) two clusters and merge them into one cluster.
4. Update the proximity matrix and compute the distances between the new cluster and each of the old clusters.
5. Repeat steps 3 and 4 until all data points are in a single cluster of size N.

Step 4 can be done in different ways. These ways lead to different hierarchical algorithms. Most of hierarchical clustering algorithms belonging to this category differ only in their definition of distances (or similarities) between clusters.

The results of these algorithms can be visualized by dendrograms. The dendrogram is a tree like diagram that shows and records the sequences of merges or splits as shown in Figure 2.2. Leaves of dendrogram consist

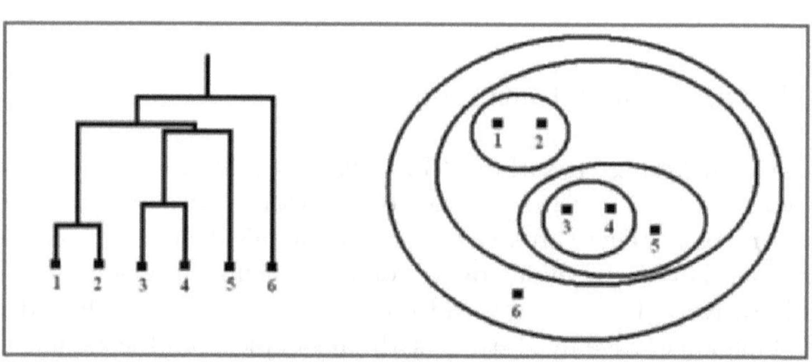

Fig. 2.2. Left: dendrogram corresponding to the clusters, right

of one object and each object is in one cluster, at each level. All objects form one cluster at the root of dendrogram. Internal nodes represent clusters constructed by merging the clusters of children.

At any stage, the calculation of distance between two clusters is based on dissimilarity between the samples from the two clusters.

Let **A** and **B** be two clusters; to calculate the distance between them we have the following methods:

1. Single-Linkage (or Nearest Neighbor):
   The distance between **A** and **B** is the shortest distance from any member of cluster **A** to any member of cluster **B**.
   $\mathbf{d}_S(\mathbf{A}, \mathbf{B}) = \min_{i \in \mathbf{A}, j \in \mathbf{B}} d_{ij}$
2. Complete-Linkage (or Furthest Neighbor):
   The distance between **A** and **B** is the largest distance from any member of cluster **A** to any member of cluster **B**.
   $\mathbf{d}_C(\mathbf{A}, \mathbf{B}) = \max_{i \in \mathbf{A}, j \in \mathbf{B}} d_{ij}$
3. Average-Linkage:
   The distance between **A** and **B** is the average of all distances between each element in cluster **A** and each element in cluster **B**.
   $\mathbf{d}_A(\mathbf{A}, \mathbf{B}) = \frac{1}{N_A N_B} \sum_{i \in \mathbf{A}} \sum_{j \in \mathbf{B}} d_{ij}$
4. Centroid:
   The distance between **A** and **B** is the distance from the centroid of cluster **A** to the centroid of cluster **B**.
5. Medoid :
   The distance between **A** and **B** is the distance between the medoids of the clusters **A** and **B**. The medoid is the most centrally located object in a cluster.

AGNES (Agglomerative Nesting) , an example agglomerative method, is a hierarchical method introduced in [134]. The AGNES algorithm uses the single-linkage method and a dissimilarity matrix to merge clusters that have least dissimilarity. DIANA (Divisive Analysis), an example of a divisive method, is also introduced in [134]. It performs in inverse order of AGNES.

2.2.1 Single-Linkage versus Complete-Linkage

Consider we have a set of 4 data points, $\mathbf{x}_1$, $\mathbf{x}_2$, $\mathbf{x}_3$ and $\mathbf{x}_4$, located as shown in Figure 2.3, and the dissimilarity matrix is shown in Table 2.1.

Table 2.1. Dissimilarity matrix for Figure 2.3

	$\mathbf{x}_2$	$\mathbf{x}_3$	$\mathbf{x}_4$
$\mathbf{x}_1$	2	5	6.4
$\mathbf{x}_2$		3	5
$\mathbf{x}_3$			4

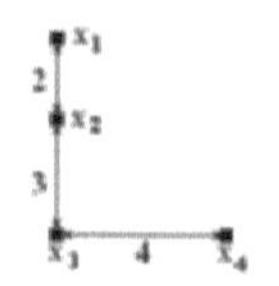

Fig. 2.3. Data set consisting of 4 data points

Using the agglomerative method with single-linkage gives the results shown in Figure 2.4, so after the last step all the data points are merged together into one cluster. Figure 2.5 shows the dendrogram to visualize the clusters.

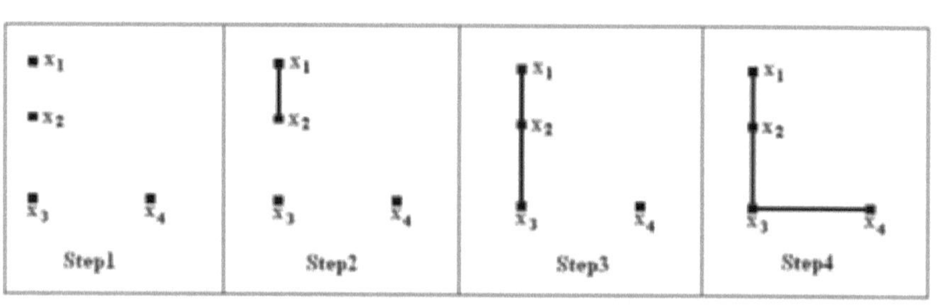

Fig. 2.4. Agglomerative method with single-linkage criterion

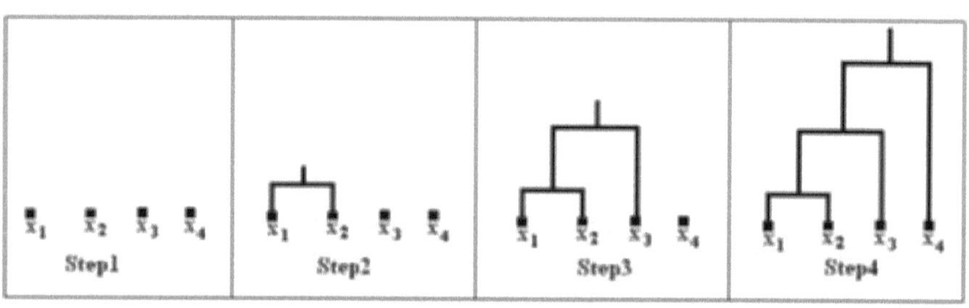

Fig. 2.5. The dendrogram associated with Figure 2.4

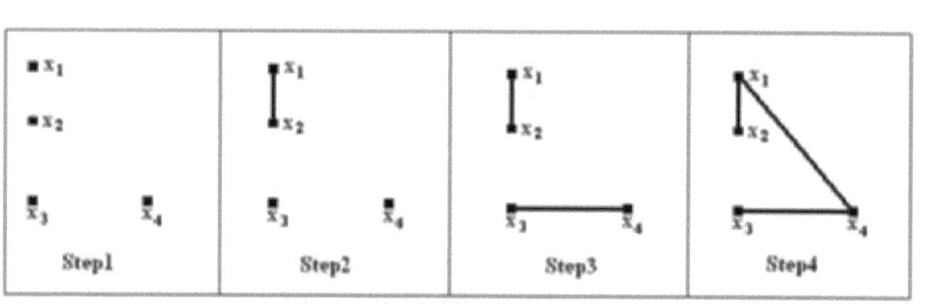

Fig. 2.6. Agglomerative method with complete-linkage criterion

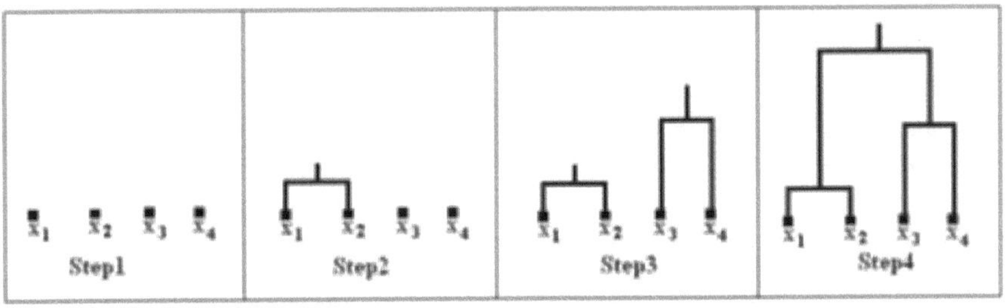

Fig. 2.7. The dendrogram associated with Figure 2.6

Figure 2.6 shows the result after each step using agglomerative method with complete-linkage. Figure 2.7 shows the dendrogram to visualize the clusters.

Strength and Limitations

The single-linkage method can handle non-elliptical shapes, but it suffers from sensitivity to noise and outliers. The complete-linkage method has less sensitivity to noise, but it tends to break large clusters and is biased towards globular clusters. Groups tend to be of equal size in terms of volume of space occupied. Average-linkage is a compromise between single-linkage and complete-linkage [214].

Hierarchical clustering methods are suitable for domains with natural nested relationships between clusters as in biology. For example, the animal kingdom can be viewed as a hierarchy of clusters. The hierarchical clustering algorithms have the following limitations:

1. Space complexity of the algorithm is $O(N^2)$, where N is the number of data samples. This space is required to store the proximity matrix.
2. Space complexity of the dendrogram $O(kN)$, where k is the number of levels.
3. Time complexity is $O(N^3)$. To complete the dendrogram till all samples are merged together into one cluster at the root, we have N steps. At each step we need to search and update the NxN proximity matrix. This time complexity can be reduced for some approaches.
4. It assumes all samples are present, so they are not incremental methods.
5. Once a decision is made to merge two clusters, it can not be undone. It is not possible to undo what had been done previously.
6. Different hierarchical clustering methods may suffer from one or more of the following limitations:
   a) Sensitivity to noise and outliers.
   b) Difficulty of handling clusters of different sizes.
   c) Breaking large clusters unnecessarily.
7. No performance function (or objective function) exists.

2.3 Partitioning Clustering

Partitioning clustering algorithms divide a given data set into clusters, so that each data point is in exactly one cluster. Given a data set of N data points $D = (\mathbf{x}_1, ..., \mathbf{x}_N)$, the task is to find K clusters $C = (\pi_1, ..., \pi_K)$, so that each data point is assigned to a unique cluster π_k. The partitioning algorithms usually attempt to minimize or maximize a performance function (or objective function) defined for clustering. There are a great many algorithms in the literature [108, 156, 214] and here we list a few of the major ones:

2.3.1 K-Means Clustering Algorithm

The K-means algorithm, first developed four decades ago [158], is one of the most popular centre-based algorithms that attempts to find K clusters which minimise the mean squared quantization error, MSQE. The algorithm tries to locate K prototypes (centroids) throughout a data set in such a way that the K prototypes in some way best represent the data. We can summarize the K-means algorithm through the following steps:

1. Initialization
   a) Define the number of prototypes (K).
   b) Designate a prototype (a vector quantity that is of the same dimensionality as the data) for each cluster.
2. Assign each data point to the closest prototype. That data point is now a member of the class identified by that prototype.
3. Calculate the new position for each prototype (by calculating the mean of all the members of that class).
4. Observe the new prototypes' positions. If these values have not significantly changed over a certain number of iterations, exit the algorithm. If they have, go back to Step 2.

The main problem of the K-means algorithm is its dependency on the prototypes' initialization . If the initial prototypes aren't chosen carefully the computation will run the chance of converging to a local minimum rather than the global minimum solution. Thus initializing prototypes appropriately can have a big effect on K-means. Figure 2.8 shows an artificial data set consisting of 3 clusters of red '*'s, and 3 prototypes with blue 'o's. In Figure 2.8, we have two different initializations of the prototypes. From Figure 2.9, the K-means algorithm gives different solutions for the same artificial data set due to its sensitivity to the prototypes' initialization.

The performance function for K-means may be written as

$$J_{Km} = \sum_{i=1}^{N} \min_{j=1}^{K} \| \mathbf{x}_i - \mathbf{m}_j \|^2 \tag{2.6}$$

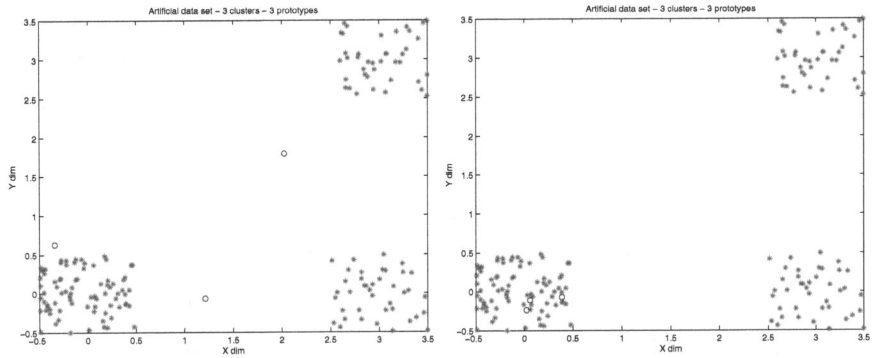

Fig. 2.8. Artificial data set with two different prototypes' initialization

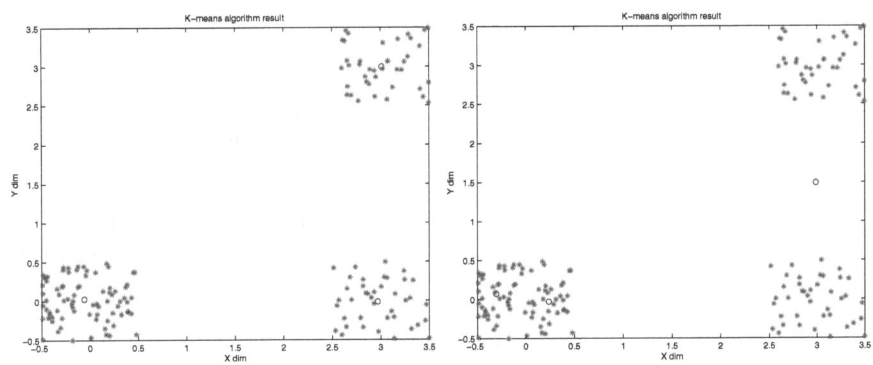

Fig. 2.9. K-means algorithm results

Note that (2.6) detects only the prototypes closest to data points and then distributes them to give the minimum performance which determines the clustering. Any prototype which is still far from data may not be utilised and does not enter any calculation that gives minimum performance. This may result in dead prototypes , which are never appropriate for any cluster. Thus initializing prototypes appropriately can have a big effect in K-means.

2.3.2 K-Means++ Clustering Algorithm

Arthur and Vassilvitskii [9] improved the K-means algorithm by substituting the random allocation of the prototypes with a seeding technique. At any given time, let $d_E^*(\mathbf{x})$ denote the shortest distance from a data point $\mathbf{x}$ to the closest prototype already chosen, then the K-means++ is as follows:

1. Choose an initial prototype $\mathbf{m}_1$ at random from the data set $\mathbf{X}$.
2. Choose the next prototype $\mathbf{m}_k$, selecting $\mathbf{m}_k = \mathbf{x}\prime \in \mathbf{X}$ with probability

$$\frac{d_E^*(\mathbf{x}\prime)^2}{\sum_{\mathbf{x}\in\mathbf{X}} d_E^*(\mathbf{x})^2} \tag{2.7}$$

3. Repeat step 2 until all prototypes have been chosen.
4. Apply the K-means clustering algorithm.

[9] gives experimental results that show the advantage of this algorithm in time and accuracy.

2.3.3 K-Medoids (or Partition Around Medoids - PAM) Clustering Algorithm

The K-medoids (or Partition Around Medoids - PAM) clustering algorithm is introduced in [134]. Each cluster is represented by one of the objects in the cluster; this object is called the medoid. The medoid of a cluster, is the most centrally located object in a cluster. The main idea of the PAM algorithm is to replace the medoid object by a non-medoid object that gives the most error reduction. This process is repeated iteratively until no changes take place. We can summarize the K-medoids algorithm through the following steps [134]:

1. Define the number of clusters (K).
2. Select K representative objects randomly to represent clusters. Let $\mathbf{i}$ represent a selected representative object (medoid), and $\mathbf{h}$ represent the non-selected object.
3. For each pair of non-selected objects, $\mathbf{h}$, and selected objects, $\mathbf{i}$:
   a) Calculate the total swapping cost ($\mathbf{TC_{ih}}$).
   b) If the total swapping cost is less than 0, ($\mathbf{TC_{ih}} < \mathbf{0}$), then $\mathbf{i}$ is replaced by $\mathbf{h}$.
   c) Assign each non-selected object to the most similar representative object.
4. Repeat step 3 until convergence.

To check if non-selected point, $\mathbf{h}$, is a better centre than the selected object, $\mathbf{i}$, we need to calculate the total swapping cost for the pair of non-selected objects, $\mathbf{h}$, and selected objects, $\mathbf{i}$. The total swapping cost for $\mathbf{i}$ and $\mathbf{h}$ equals the sum of swapping cost on each non-centre point:

$$\mathbf{TC}_{ih} = \sum_j \mathbf{C}_{jih} \tag{2.8}$$

The swapping cost of the non-centre point $\mathbf{j}$ equals to the distance between $\mathbf{j}$ and its closest centre after swapping $\mathbf{i}$ and $\mathbf{h}$ minus the distance between $\mathbf{j}$ and its closest centre before swapping $\mathbf{i}$ and $\mathbf{h}$, see Figure 2.10. The PAM algorithm has less sensitivity than K-means algorithm to the noise and outliers as a mediod is less influenced by outliers than a mean [214]. The PAM algorithm works effectively for small data sets, but doesn't scale well for large data sets.

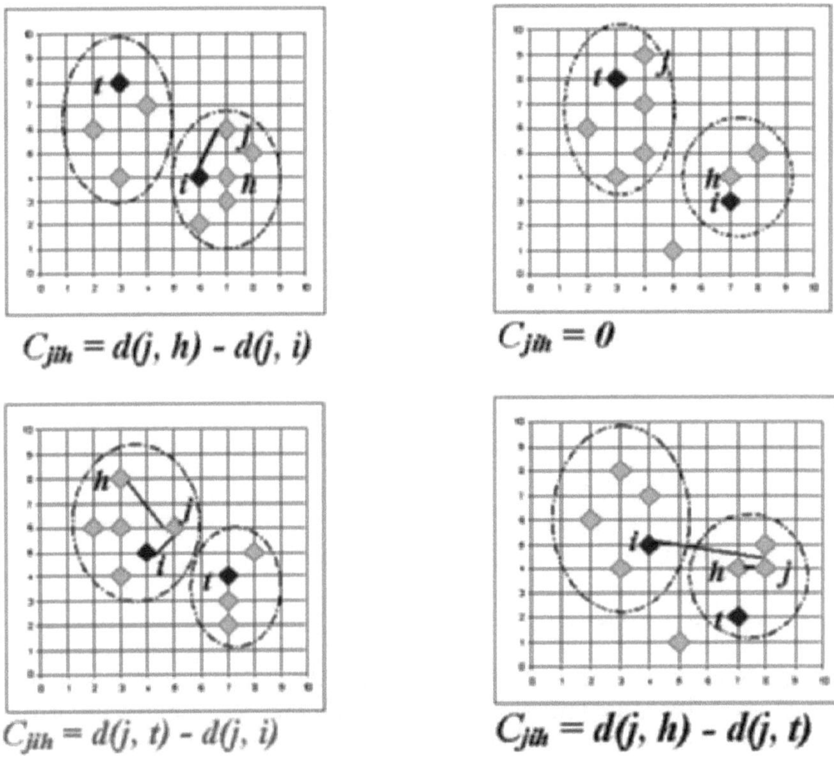

Fig. 2.10. Swapping cost on non-centre point **j** [108, 48]

2.3.4 Fuzzy C-Means Clustering Algorithm (FCM)

FCM algorithm was developed by Dunn in 1973 [71] and improved by Bezdek in 1981 [32]. It is frequently used in pattern recognition. FCM algorithm allows one data point to belong to two or more clusters. The performance function of this algorithm is:

$$J_{FK} = \sum_{i=1}^{N} \sum_{k=1}^{K} w_{ik}^{P} d^2(\mathbf{x}_i, \mathbf{m}_k). \tag{2.9}$$

subject to the constraint $\sum_{j=1}^{K} w_{ij} = 1$.

Where $P > 1$ is a fuzzification (or weighting) exponent that controls how fuzzy the result will be.

w_{ij} describes the degree of membership of data sample $\mathbf{x}_i$ with the prototype $\mathbf{m}_j$.

$d(a, b)$ is the Euclidean distance between a and b.

To minimize the performance function (2.9), the prototypes are allocated according to [32, 31]:

$$\mathbf{m}_k = \frac{\sum_{i=1}^{N} w_{ki}^{P} \mathbf{x}_i}{\sum_{i=1}^{N} w_{ki}^{P}} \tag{2.10}$$

$$\text{where e.g. } w_{ki} = \frac{1}{\sum_{q=1}^{K} \left(\frac{d^2(\mathbf{x}_i, \mathbf{m}_k)}{d^2(\mathbf{x}_i, \mathbf{m}_q)} \right)^{\frac{1}{P-1}}} \tag{2.11}$$

Fuzzy C-means clustering algorithm (FCM) has the following steps:

1. Initialization: initialize K (or C) prototypes randomly.
2. Calculate the weights using (2.11).
3. Update the prototypes using (2.10).
4. Repeat steps 2 and 3 until the changes of the weights become less than a specified threshold ϵ.

The disadvantages of FCM are its sensitivity to noise, outliers and initial conditions, and it requires a long computational time [204].

2.3.5 Soft K-Means Clustering Algorithm

The soft K-means algorithm is discussed in [156]. It is a centre-based algorithm; K prototypes are initialized randomly and then are updated iteratively in a way to identify clusters. The prototypes are updated according to

$$\mathbf{m}_k = \frac{\sum_n r_{kn} \mathbf{x}_n}{\sum_{j,n} r_{jn}} \tag{2.12}$$

$$\text{where e.g. } r_{kn} = \frac{\exp(-\beta \parallel \mathbf{x}_n - \mathbf{m}_k \parallel)}{\sum_j \exp(-\beta \parallel \mathbf{x}_n - \mathbf{m}_j \parallel)} \tag{2.13}$$

There are still problems with soft K-means: we illustrate by applying the soft K-means algorithm to the artificial data sample shown in Figure 2.11 in which we deliberately choose a poor initialisation. We find that with soft K-means it is important to choose a good value for β; if we choose a poor value we may have poor results in finding the clusters as shown in Figure 2.12. However, even with a good value, we often still find that the soft K-means has the problem of sensitivity to the prototypes' initialization. As shown in Figure 2.13, while the soft K-means succeeds in identifying the clusters (right diagram), sometimes it failed for the same data sample when we used a different initialization of the prototypes (left diagram).

2.3.6 K-Harmonic Means Clustering Algorithm (KHM)

KHM is a centre-based clustering algorithm which uses the harmonic averages of the distances from each data point to the prototypes as components to its performance function. The performance function of KHM is

$$J_{HA} = \sum_{i=1}^{N} \frac{K}{\sum_{k=1}^{K} \frac{1}{\|\mathbf{x}_i - \mathbf{m}_k\|^2}} \tag{2.14}$$

To minimize the performance function differentiate

$$\frac{\partial J_{HA}}{\partial \mathbf{m}_k} = -K \sum_{i=1}^{N} \frac{2(\mathbf{x}_i - \mathbf{m}_k)}{\| \mathbf{x}_i - \mathbf{m}_k \|^4 \{\sum_{l=1}^{K} \frac{1}{\|\mathbf{x}_i - \mathbf{m}_l\|^2}\}^2} \tag{2.15}$$

Setting this equal to 0 and solving for all $\mathbf{m}_k$'s, we get a recursive formula

$$\mathbf{m}_k^{(new)} = \frac{\sum_{i=1}^{N} \frac{1}{\|\mathbf{x}_i - \mathbf{m}_k\|^4 (\sum_{l=1}^{K} \frac{1}{\|\mathbf{x}_i - \mathbf{m}_l\|^2})^2} \mathbf{x}_i}{\sum_{i=1}^{N} \frac{1}{\|\mathbf{x}_i - \mathbf{m}_k\|^4 (\sum_{l=1}^{K} \frac{1}{\|\mathbf{x}_i - \mathbf{m}_l\|^2})^2}} \tag{2.16}$$

There are some practical issues, division by zero, to deal with the implementation details of which are given in [240, 244]. Zhang subsequently developed a generalized version of the KHM [241, 242, 243].

[241, 244] have extensive simulations showing that this algorithm converges to a better solution (less prone to finding a local minimum because of poor initialization) than both standard K-means or a mixture of experts trained using the EM algorithm.

2.3.7 Kernel K-Means Clustering Algorithm

A number of kernel methods have been proposed in [43, 88, 212] to increase the separability of clusters. The kernel K-means algorithm was proposed [67, 197]

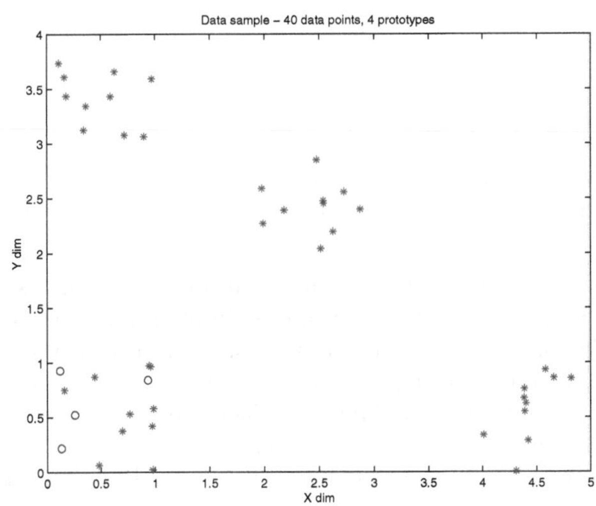

Fig. 2.11. Data set is shown as 4 clusters of red '*'s; initial positions of the prototypes are shown as blue 'o's

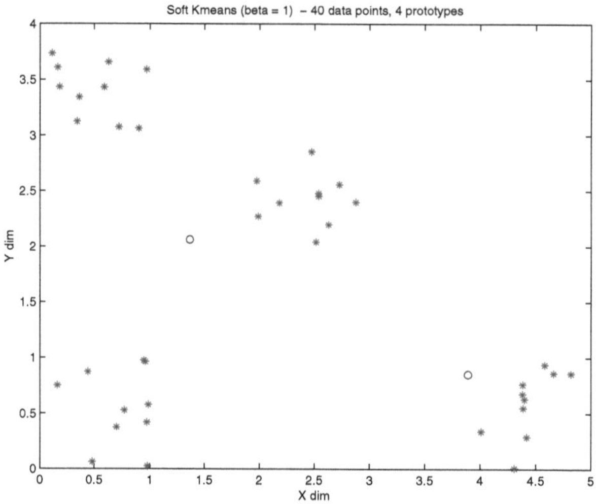

Fig. 2.12. Due to poor initialisation, the soft K-means algorithm has failed to identify the four clusters, some prototypes are joint together

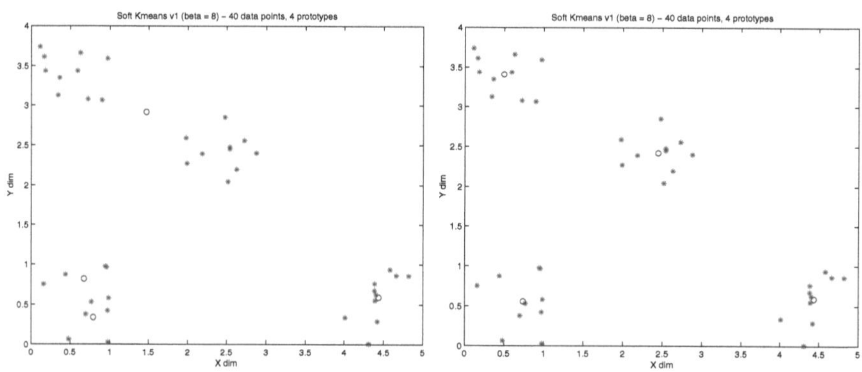

Fig. 2.13. The vagaries of simulations: left, the soft K-means failed to identify the four clusters but on the right is successful with the same data set

recently as an extension of the standard K-means algorithm. Usually the extension from K-means to kernel K-means is simply realised by expressing the distance in the form of a kernel function [103, 173]. In the kernel K-means algorithm all data points are mapped, before clustering, to a higher dimensional feature space by using a kernel function. Then the K-means algorithm is applied in the new feature space to identify clusters.

The performance function of the kernel K-means algorithm can be written as:

$$J_{KK} = \sum_{i=1}^{N} \min_{k=1}^{K} \parallel \phi(\mathbf{x}_i) - \mathbf{m}_k \parallel^2 \tag{2.17}$$

where $\mathbf{m}_k = \frac{\sum_{\mathbf{x}_i \in \pi_k} \phi(\mathbf{x}_i)}{|\pi_k|}$, π_k represents the kth cluster, $|\pi_k|$ represents the number of points in π_k and $\phi(\mathbf{x}_i)$ represents the data point $\mathbf{x}_i$ after mapping to the kernel space.

By using the inner products to compute the Euclidean distances between data points and prototypes

$$\| \phi(\mathbf{x}_i) - \mathbf{m}_k \|^2 = \phi(\mathbf{x}_i).\phi(\mathbf{x}_i) - \frac{2\sum_{\mathbf{x}_j \in \pi_k} \phi(\mathbf{x}_i).\phi(\mathbf{x}_j)}{|\pi_k|} + \frac{\sum_{\mathbf{x}_j, \mathbf{x}_l \in \pi_k} \phi(\mathbf{x}_j).\phi(\mathbf{x}_l)}{|\pi_k|^2}.$$

The dot products $\phi(\mathbf{x}).\phi(\mathbf{m})$ are computed using kernel function $kernel(\mathbf{x}, \mathbf{m})$, see Table 2.2 for commonly used kernel functions

$$\| \phi(\mathbf{x}_i) - \mathbf{m}_k \|^2 = kernel(\mathbf{x}_i, \mathbf{x}_i) - \frac{2\sum_{\mathbf{x}_j \in \pi_k} kernel(\mathbf{x}_i, \mathbf{x}_j)}{|\pi_k|} + \frac{\sum_{\mathbf{x}_j, \mathbf{x}_l \in \pi_k} kernel(\mathbf{x}_j, \mathbf{x}_l)}{|\pi_k|^2}.$$

So, in feature space we can calculate the distances between mapped data points and prototypes without knowing explicit representations or coordinates of $\phi(\mathbf{x}_i)$. We compute the kernel matrix between all data points by using a kernel function.

Table 2.2. Popular kernel functions

Polynomial Kernel	$kernel(\mathbf{a}, \mathbf{b}) = (\mathbf{a}.\mathbf{b} + c)^d$
Gaussian Kernel	$kernel(\mathbf{a}, \mathbf{b}) = \exp(- \| \mathbf{a} - \mathbf{b} \|^2 / 2\sigma^2)$
Sigmoid Kernel	$kernel(\mathbf{a}, \mathbf{b}) = \tanh(c(\mathbf{a}.\mathbf{b}) + \theta)$

Kernel K-means algorithm steps are:

1. Initialize K partitions (clusters); assign each data point randomly to a cluster, π_k.
2. Compute the kernel matrix between all data points.
3. Update assignments, assign each data point to a cluster π_k
   $k = \arg\min_i \left[-\frac{2}{|\pi_i|} \sum_{\mathbf{x}' \in \pi_i} kernel(\mathbf{x}, \mathbf{x}') + \frac{1}{|\pi_i|^2} \sum_{\mathbf{x}', \mathbf{x}'' \in \pi_i} kernel(\mathbf{x}', \mathbf{x}'') \right]$
4. Repeat step 3 until convergence.

2.3.8 Spectral Clustering Algorithm

Spectral clustering methods [154, 202, 174] use the top eigenvectors of an affinity matrix , derived from similarities between data points, to obtain a clustering of the data. Several methods have been proposed [131, 174], each of which uses the eigenvectors in slightly different ways. The popular objective (or performance) function used in spectral clustering is to minimize the normalized cut [203, 238].

Consider we have a data set $\mathbf{X} = \mathbf{x}_1, ..., \mathbf{x}_N$ which we want to cluster into K clusters. Then the main steps for spectral algorithms are as follows:

1. Construct the affinity (or similarity) matrix $\mathbf{S}$ which measures the weights (or similarities) between all data points e.g. $\mathbf{s}_{ij} = \exp(- \parallel \mathbf{x}_i - \mathbf{x}_j \parallel /2\sigma^2)$ if $i \neq j$, and $\mathbf{s}_{ii} = 0$.
2. Compute the diagonal matrix $\mathbf{D}$, where D_{ii} = sum of ith row in $\mathbf{S}$.
3. Construct the matrix $\mathbf{L} = \mathbf{D}^{-1/2}\mathbf{S}\mathbf{D}^{-1/2}$.
4. Compute the K eigenvectors $\mathbf{u}_1, ..., \mathbf{u}_K$ of $\mathbf{L}$ with largest eigenvalues.
5. Construct the matrix $\mathbf{U} = [\mathbf{u}_1, ...\mathbf{u}_K] \in \Re^{N \times K}$, where $\mathbf{U}$ contains the eigenvectors as columns.
6. Form the matrix $\mathbf{T} \in \Re^{N \times K}$ from $\mathbf{U}$ by normalizing each of $\mathbf{U}$'s rows to have unit length.
7. Treat each row of $\mathbf{T}$ as a point in $\Re^K$ and then cluster them into K clusters using K-means algorithm.
8. Assign the original point $\mathbf{x}_i$ to cluster k if row i, $\mathbf{t}_i$, of the matrix $\mathbf{T}$ was assigned to cluster k.

The main trick of spectral clustering algorithms is to change the representation of the original data points $\mathbf{x}_i$ to points $\mathbf{t}_i \in \Re^K$ by using the spectrum of the affinity matrix.

2.4 Topology Preserving Mapping

Humans are good at analyzing data in 2D or 3D. Most scientific data sets that we need to deal with are high dimensional data. It would help if we could visualize structure of the data in 2D or 3D. In this section we show some recent and famous methods that can be used for dimensionality reduction and visualization. The topology preserving mapping can be defined as a mapping that preserves neighborhood relations between input vectors, so nearby neurons are mapped from similar samples.

2.4.1 Self-Organizing Map (SOM)

The Self-Organizing Map, developed by Teuvo Kohonen [137, 138], is one of the most popular neural network models. It belongs to the category of competitive learning networks and is based on unsupervised learning , which means that no human intervention is needed during the learning process. The algorithm itself is able to find interesting features of the data without pre-knowledge about it. As shown in Figure 2.14, the map unit is the output layer, usually a two-dimensional grid of nodes, neurons, that is fully connected to the input layer. Each neuron is a node containing a template against which input patterns are matched. All nodes are presented with the same input pattern in parallel and compute the distance between their template and the input in parallel. Only the node with the closest match between the input

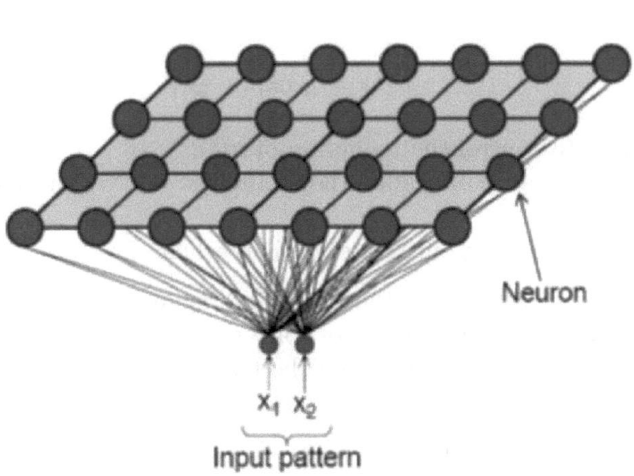

Fig. 2.14. A self-organising network arranged in a rectangular topology

and its template produces an active output. The disposition of the nodes (or neurons) in that grid can be different, e.g. sheet, cylinder, toroid etc.

One of the main properties of the SOM is that it provides a topology preserving mapping from a high dimensional space to map units [75, 76, 190]. The property of topology preservation means that the points that are closer to each other in output space are mapped from points that are close together in data space. The SOM can thus serve as a cluster analyzing tool of high-dimensional data. Though some close points in data space may map to distant points in output space.

After training the SOM, a vector is presented to the input layer, and the node whose weight vector is most similar to this input vector will be activated.

The SOM algorithm can be summarised as follows:

1. Initialize the prototypes $\mathbf{m}_k$, $k = 1, 2, ..., K$ (codebook vectors).
2. Select at random an input point $\mathbf{x}_i$ (input pattern)
   a) Search for the nearest prototype $\mathbf{m}_{k\prime}$. The nearest prototype, the winning neuron, is the closest neuron to the presented sample and is called the Best Matching Unit (BMU). This can be done by using the Euclidean distance.
   b) Adjust the BMU and its neighbours so that they become more similar to the presented sample by using

$$\mathbf{m}_k(t+1) = \mathbf{m}_k(t) + \eta(t)(\mathbf{x}_i(t) - \mathbf{m}_k(t))\Lambda_{k,k\prime}(t) \tag{2.18}$$

where $\eta(t)$ is the learning rate, $\mathbf{x}_i(t)$ is the presented sample, and $\Lambda_{k,k\prime}(t)$ is often a monotonically decreasing function of the distance between k and $k\prime$, known as the neighborhood function.

$$\Lambda_{k,k\prime}(t) = \exp\left(\frac{- \parallel \mathbf{r}_k - \mathbf{r}_{k\prime} \parallel}{2\sigma^2(t)}\right) \tag{2.19}$$

where

$$\sigma(t) = \sigma_o \exp(-\frac{t}{T_1}) \tag{2.20}$$

T_1 is used to control the reduction of $\sigma(t)$.

c) Decrease the range of neighborhood and the value of the learning rate a little.

3. Repeat 2 until a convergence criterion is reached.

There are some drawbacks to the SOM algorithm [81, 180] like the lack of an objective function and no proof of convergence. However, it is a widely used technique [132, 179], that has been the basis for many algorithms [141].

2.4.2 Generative Topographic Map (GTM)

The Generative Topographic Map (GTM) [34, 35, 36, 213, 218] is a mixture of experts model which treats the data as having been generated by a set of latent points. The GTM can be described as a non-linear latent variable model that defines a mapping from the latent space to the data space, generating a probability density within the latter. It can be seen as a probabilistic alternative to the Self-Organizing Map (SOM) resolving many of its associated theoretical problems. Since the GTM is non-linear, the relationship between data and its visual representation may be far from trivial, but a better understanding of this relationship can be gained by computing the magnification factor. As shown in Figure 2.15, the GTM structure consists of a 2 or 3 dimensional grid of latent points (or nodes), in latent space.

These latent points are mapped non-linearly from latent space into data space by using a fixed grid of non-linear basis functions and a set of adjustable weights $\mathbf{m}(\mathbf{t};\mathbf{W})$. This mapping defines a set of reference vectors (or prototypes) embedded in the data space. The non-linear function $\mathbf{m}(\mathbf{t};\mathbf{W})$

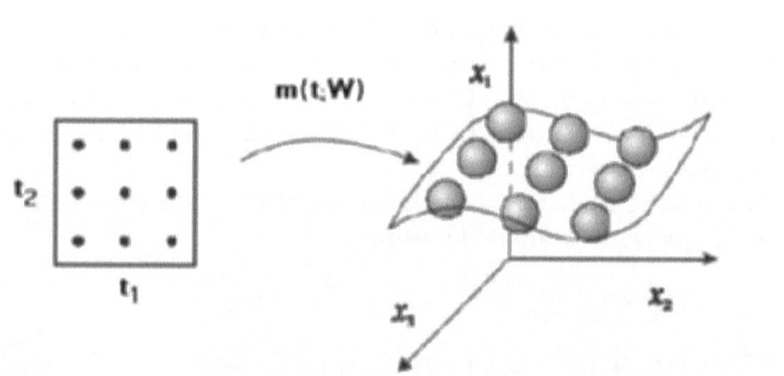

Fig. 2.15. GTM: The latent points are projected non linearly onto the data space by using non linear basis functions and adjustable weights [36]

maps every latent point $\mathbf{t}_k \in \mathbf{R}^q$ to a prototype $\mathbf{m}_k \in \mathbf{R}^d$ embedded into the data space, where normally $q < d$.

$$\mathbf{m}(\mathbf{t};\mathbf{W}) = \Phi(\mathbf{t})\mathbf{W} \tag{2.21}$$

where the elements of $\Phi(\mathbf{t})$ consist of M fixed basis functions, $\Phi_i(\mathbf{t})$, and $\mathbf{W}$ is a $M \times d$ matrix.

Every prototype $\mathbf{m}_k$ forms the centre of an isotropic Gaussian distribution in data space

$$p(\mathbf{x}|\mathbf{t}_k, \mathbf{W}, \beta) = \left(\frac{\beta}{2\pi}\right)^{\frac{d}{2}} \exp\left(-\frac{\beta}{2} \parallel \mathbf{x} - \mathbf{m}_k \parallel^2\right) \tag{2.22}$$

where $\mathbf{x}$ is a point in the data space and β denotes the noise inverse variance. The probability distribution in the data space for GTM model is given by:

$$p(\mathbf{x}|\mathbf{W}, \beta) = \int p(\mathbf{x}|\mathbf{t}, \mathbf{W}, \beta)p(\mathbf{t})d\mathbf{t}. \tag{2.23}$$

Choosing $p(\mathbf{t})$ as a set of K equally weighted delta functions on a regular grid,

$$p(\mathbf{t}) = \frac{1}{K}\sum_{k=1}^{K} \delta(\mathbf{t} - \mathbf{t}_k) \tag{2.24}$$

the integral in (2.23) turns into a sum,

$$p(\mathbf{x}|\mathbf{W}, \beta) = \frac{1}{K}\sum_{k=1}^{K} \left(\frac{\beta}{2\pi}\right)^{\frac{d}{2}} \exp\left(-\frac{\beta}{2} \parallel \mathbf{x} - \mathbf{m}_k \parallel^2\right) \tag{2.25}$$

From equation (2.25), the GTM defines a constrained mixture of Gaussians [123, 129], since the centres of the mixture components can not move independently of each other; they depend on the mapping $\mathbf{m}(\mathbf{t};\mathbf{W})$. Moreover, all components of the mixture share the same variance, and the mixing coefficients are all fixed at $\frac{1}{K}$. The corresponding log-likelihood is defined as

$$L(\mathbf{W}, \beta) = \sum_{i=1}^{N} \ln[p(\mathbf{x}_i|\mathbf{W}, \beta)] \tag{2.26}$$

The Expectation Maximisation algorithm (EM) can be used to maximize (2.26) with respect to the parameters of the mixture, $\mathbf{W}$ and β.

After training and convergence, we can compute the corresponding posterior distribution in latent space for any given point $\mathbf{x}$ in data space by

$$p(\mathbf{t}_k|\mathbf{x}) = \frac{p(\mathbf{x}|\mathbf{t}_k, \mathbf{W}, \beta)p(\mathbf{t}_k)}{\sum_{k'=1}^{K} p(\mathbf{x}|\mathbf{t}_{k'}, \mathbf{W}, \beta)p(\mathbf{t}_{k'})} \tag{2.27}$$

Data can be visualized by projecting each data point onto the latent space to reveal interesting features. There are two ways to visualize the whole data set. For each data point $\mathbf{x}_i$, we can use:

1. The mode of the posterior distribution (posterior-mode projection)

$$\mathbf{t}_k^{mode} = \arg\max_{\mathbf{t}_k} p(\mathbf{t}_k|\mathbf{x}_i) \tag{2.28}$$

2. The mean of the posterior distribution (posterior-mean projection)

$$\mathbf{t}_k^{mean} = \sum_{k=1}^{K} \mathbf{t}_k p(\mathbf{t}_k|\mathbf{x}_i) \tag{2.29}$$

There are two principal limitations of the basic GTM model. The computational effort required will grow exponentially with the intrinsic dimensionality of the density model. However, if the intended application is visualization, this will typically not be a problem. The other limitation is that the initialization of the parameters can lead the algorithm to a local minimum [34].

2.4.3 Topographic Product of Experts (ToPoE)

Fyfe [93] has extended the Product of Gaussian model by allowing latent points to have different responsibilities depending on the data point presented:

$$p(\mathbf{x}_i|\Theta) \propto \Pi_{k=1}^{K} \left(\frac{\beta}{2\pi}\right)^{\frac{d}{2}} \exp\left(-\frac{\beta}{2} \parallel \mathbf{x} - \mathbf{m}_k \parallel^2 r_{ik}\right) \tag{2.30}$$

where r_{ik} is the responsibility of the kth expert for the data point $\mathbf{x}_i$. Thus all the experts are acting in concert to create the data points but some will take more responsibility than others. ToPoE has the same structure as the GTM, with a number of latent points that are mapped to a feature space by M Gaussian functions, and then into the data space by a matrix $\mathbf{W}$. Each latent point $\mathbf{t}$ indexed by k is mapped, through a set of M fixed basis functions $\Phi_1(\mathbf{t})$, $\Phi_2(\mathbf{t})$,..., $\Phi_M(\mathbf{t})$, to a prototype in data space $\mathbf{m}_k = \mathbf{\Phi}(\mathbf{t}_k)\mathbf{W}$. But the weights of the mapping are trained in a different way based on a product of experts in order to maximise the probability of the data under this model. The major advantage which a product of experts method has is that it is possible to get very much sharper probability density functions with a product than with a sum of experts.

ToPoE algorithm steps are

1. Initialize the weights $\mathbf{W}$ randomly and spread the centres of the M basis functions uniformly in latent space.
2. Initialize the K latent points uniformly in latent space.
3. Initialize counter $= 0$.
4. Compute $\mathbf{m}_k = \mathbf{\Phi}(\mathbf{t}_k)\mathbf{W}$.
5. Select a random data point $\mathbf{x}_i$.

6. Calculate the responsibilities that the kth latent point has for the ith data point
$$r_{ik} = \frac{\exp\left(-\gamma \parallel \mathbf{x}_i - \mathbf{m}_k \parallel^2\right)}{\sum_{k'=1}^{K} \exp\left(-\gamma \parallel \mathbf{x}_i - \mathbf{m}_{k'} \parallel^2\right)} \tag{2.31}$$
If no prototypes are close to the data point, then $r_{ik} = \frac{1}{K}, \forall k$.
7. Update $\mathbf{W}$
$$\Delta_i w_{md} = \sum_{k=1}^{K} \eta \phi_{km} (x_i^{(d)} - m_k^{(d)}) r_{ik} \tag{2.32}$$
where $x_i^{(d)}$ is the dth coordinate of $\mathbf{x}_i$, and $m_k^{(d)}$ is the dth coordinate of $\mathbf{m}_k$.
8. If counter < MAXCOUNT, increment counter and return to step 4.

Equation (2.32) is derived from the minimization of $-\log(p(\mathbf{x}_i|\Theta))$ with respect to a set of parameters that generate the $\mathbf{m}_k$.

Since $-\log(p(\mathbf{x}_i|\Theta)) \propto \sum_{k=1}^{K} \parallel \mathbf{x}_i - \mathbf{m}_k \parallel r_{ik}$, the minimization of $-\log(p(\mathbf{x}_i|\Theta))$ is equal to minimizing the weighted mean square error.

2.4.4 Harmonic Topographic Mapping (HaToM)

The Harmonic Topographic Map [95, 161, 182, 184, 185, 189] was developed as an alternative to the ToPoE [93], and is also based on the GTM. The HaToM shares the structure of the GTM, with a latent space projected non-linearly through a feature space into data space by M fixed Gaussian functions and a matrix W. But the similarity ends there because the objective function is not a probabilistic function like the GTM neither it is optimised with the Expectation-Maximization (EM) algorithm. Instead, the HaToM [183, 187, 188] uses the K-Harmonic Means algorithm for learning process. Each data point is visualized as residing at the prototype on the map which would win the competition for that data point. However we can do rather better by defining the responsibility that the jth prototype has for the ith data point as

$$r_{ji} = \frac{\exp(-\gamma \parallel \mathbf{x}_i - \mathbf{w}_j \parallel^2)}{\sum_k \exp(-\gamma \parallel \mathbf{x}_i - \mathbf{w}_k \parallel^2)} \tag{2.33}$$

We then project points taking into account these responsiblities: let y_{ij} be the projection of the ith data point onto the jth dimension of the latent space; then

$$y_{ij} = \sum_k t_{kj} r_{ki} \tag{2.34}$$

where t_{kj} is the jth coordinate of the kth latent point. When we use these algorithms for visualisation purposes, it is these y-values (which are typically two dimensional coordinates) which we use. Note that this method represents each data point $\mathbf{x}_i$ by a value $\mathbf{y}_i$ where $\mathbf{y}_i$ is a weighted sum of the coordinates of the original latent points. An alternative (which is typically used the SOM) is to find the latent point with greatest responsibility for the data point and allocate its $\mathbf{y}_i$ value at this latent point.

Recently, [162, 186] introduces two versions of Hatom, D-HaToM and M-HaToM. D-HaToM (for Data-driven HaToM) takes the data-greedy properties of the K-Harmonic Means algorithm and is very good at covering the whole data space. However it will, because of this data-greed property, be very affected by outliers and indeed, in the normal course of the algorithm, it will seek out the outer edges of clusters. This is sometimes valuable but if one wishes to identify a smooth manifold, the M-HaToM (or Model-driven HaToM) is preferrable: the M-HaToM is able to find a smooth manifold with any number of latent points, while D-HaToM can not.

2.5 Conclusion

In this chapter we have reviewed techniques and methods related to the algorithms which will be developed in this book. We have concentrated on clustering algorithms in the hierarchical and partitional categories. We have reviewed some of the major standard clustering algorithms such as K-means algorithm, soft K-means algorithm, kernel K-means, etc. Also we have reviewed some of topology preserving mapping algorithms such as the self-organizing map (SOM), the generative topographic map (GTM), and the harmonic topographic mapping (HaToM). In the remainder of this book, we will develop algorithms which overcome some of the deficiencies of the existing clustering algorithms.

3

Review of Linear Projection Methods

In this chapter, we briefly review the background of projection methods.

3.1 Linear Projection Methods

Exploratory data analysis is a set of methods with which we try to extract as much information as possible from a data set of high dimension and huge volume. However, performing analysis of complex data usually involves a large number of variables and analysis with a large number of variables generally requires a large amount of memory and computational power and may generalize poorly to new samples. Many techniques change the basis of the considered data space by projecting the data to a lower dimensional space. The basic idea is to find some suitable function $\varphi : \Re^D \mapsto \Re^d, d \ll D$, which maps the original data sample $\mathbf{x} \in \Re^D$ into a d-dimensional manifold by $\varphi(\mathbf{x}) = y$, where $\mathbf{x} \in \Re^D, \mathbf{y} \in \Re^d$. In this section, we review several projection methods in detail.

3.1.1 Principal Component Analysis

Principal component analysis (PCA) is a well-known statistical technique for multivariate data analysis. The goal of principal component analysis is to find an orthogonal basis such that the elements of projection of the data in subspace $\Re^d$ become uncorrelated, thus this method focuses on the first- and second- order statistics. The variances of the projections of the data are maximized by finding a set of filters W, so that the first principal component (PC) accounts for the maximal variance based on the first filter $\mathbf{w_1}$, the second principal component in the direction othogonal to the first PC corresponds to most of the remaining variance based on $\mathbf{w_2}$, and so on. Figure 3.1 shows the first principal component of a two-dimensional data cloud and the second principal component would be orthogonal to the first.

W.A. Barbakh et al.: Non-Stand. Param. Adapt. for Explor. Data Analys., SCI 249, pp. 29–48.
springerlink.com

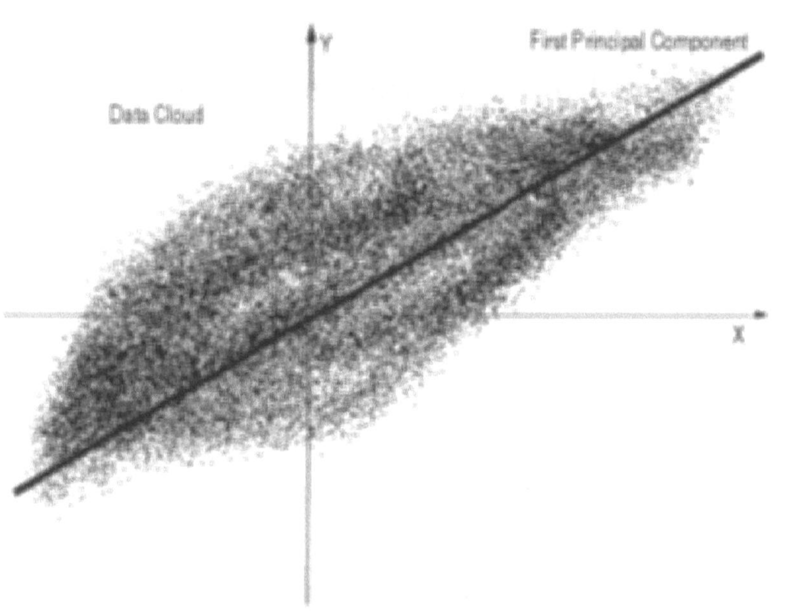

Fig. 3.1. The first principal component of a two-dimensional data cloud; the first principal component is in the direction with maximal variance

In mathematical terms, we may think of PCA as a linear combination

$$y_i = \sum_{k=1}^{D} w_{ik} x_k = \mathbf{w}_i^T \mathbf{x} \tag{3.1}$$

where x_k is the $k-th$ element of the data set and $\mathbf{w}_1 \ldots \mathbf{w}_d$ is a set of orthogonal unit norm weight vectors, or *filters*. The factor y_i represents the $i-th$ principal component, where y_1 is called the first principal component and thus the variance of y_1 is maximally large. y_i is constrained to be of maximal variance subject to y_i being uncorrelated with all the previously found principal components:

$$\mathbf{E}_{\mathbf{x}}\{y_i y_k\} = 0, k = 1, \ldots, i-1. \tag{3.2}$$

where we have assumed zero mean data. Since the variance of y_i depends on the weight vector $\mathbf{w}_i$, we look for such a weight vector maximizing the PCA criterion

$$J_i(\mathbf{w}_i) = \mathbf{E}\{y_i^2\} = \mathbf{E}\{(\mathbf{w}_i^T\mathbf{x})^2\} = \mathbf{w}_i^T\mathbf{E}\{\mathbf{x}\mathbf{x}^T\}\mathbf{w}_i = \mathbf{w}_i^T\mathbf{C}_{\mathbf{x}}\mathbf{w}_i, \tag{3.3}$$

where $||\mathbf{w}_i|| = 1$ and the matrix $\mathbf{C}_{\mathbf{x}}$ is the $D \times D$ covariance matrix of the data $\mathbf{x}$

$$\mathbf{C}_{\mathbf{x}} = \mathbf{E}\{\mathbf{x}\mathbf{x}^T\}, \tag{3.4}$$

if the data $\mathbf{x}$ is zero-mean. It has been shown [128] that the maximization occurs when the weight vectors satisfy

$$\mathbf{C}\mathbf{w}_i = \lambda_i \mathbf{w}_i. \tag{3.5}$$

where λ_i and $\mathbf{w}_i$ are the eigenvalues and eigenvectors of the covariance matrix $\mathbf{C}$. Thus, the solution of the PCA problem is given in terms of the unit-length eigenvectors $\mathbf{w}_1 \ldots \mathbf{w}_d$ of the matrix $\mathbf{C}_\mathbf{x}$, with the corresponding eigenvalues $\lambda_1 \ldots \lambda_d$ satisfying $\lambda_1 \geq \lambda_2 \geq \ldots \geq \lambda_d$, and the PCA problem can then be solved by an eigen-decomposition method.

In this book, we have more interest in the problem of PCA with on-line learning. Neural networks provide a novel way to perform on-line computation of the PCA. The most famous ones are Oja's algorithms [175, 176]. Fyfe [90, 89] implemented PCA by a negative feedback network defined by (3.6) - (3.8). The network has an N-dimensional input vector and an M-dimensional output vector, with w_{ij} being the weight linking the jth input to i^{th} output. In the forward stage, the activation passes input to the output through the weights in (3.6). Then the activation is fed back through the weights from the output and finally the weights are updated by (3.8).

$$y_i = \sum_{j=1}^{N} w_{ij} x_j \tag{3.6}$$

$$e_j = x_j - \sum_{i=1}^{M} w_{ij} y_i \tag{3.7}$$

$$\Delta w_{ij} = \eta e_j y_i \tag{3.8}$$

By using PCA, a high dimensional data set can be mapped to lower dimensional principal components with the most important linear characteristics retained and so PCA can be regarded as a useful feature extraction technique.

3.1.2 Exploratory Projection Pursuit

In the previous subsection, we can see that PCA is a powerful tool to investigate the structure of a high-dimensional data set. However, the identification of maximum variance filters is not guaranteed to find projections which reveal interesting structure in a data set. [92] has given a good example as shown in Figure 3.2, in which two data clusters are represented by two ellipsoids. The first principal component would be diagonal either from the top left to bottom right or from bottom left to top right, yet the structure of the data set is hidden in such projections.

Exploratory Projection Pursuit (EPP) (see Friedman [85], Fyfe [92]) defines a group of techniques based on projection pursuit (see Friedman [86, 87]) designed to investigate the structure in high dimensional data sets by finding "interesting" directions in the data space. The basic idea is that rather than

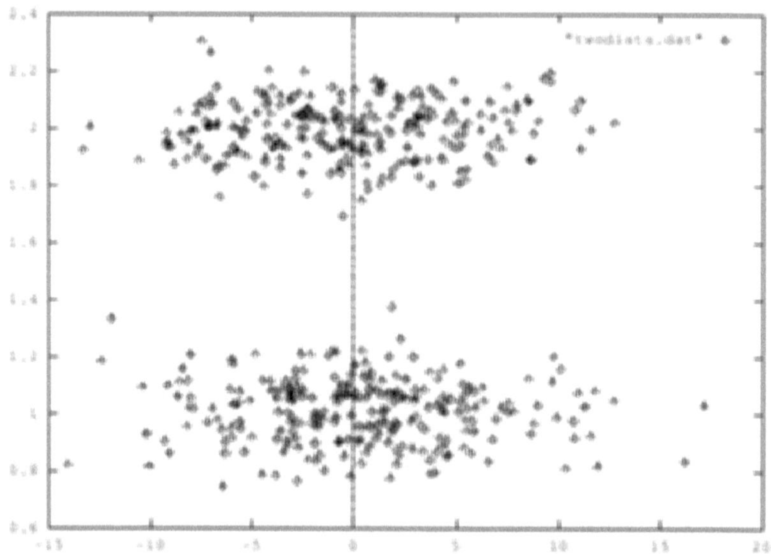

Fig. 3.2. The projections onto either principal component would hide the structure of the data set [92]

identifying the structure of a data set in high-dimensional data space, we look for the structure that is revealed in projections to a lower-dimensional subspace. Diaconis and Freedman [68] have noted that the distribution of projections of high-dimensional data tends to be a Gaussian distribution, which suggest that if we want to identify "interesting" features in data, we need to find such directions **w** on which the projections are as non-Gaussian as possible.

Therefore, we define an index function of the direction **w**, $I(\mathbf{w})$, which is maximized when the distribution of projections on **w** is furthest from Gaussian. A common way to measure non-Gaussianity is based on higher order moments of the distribution as shown in Figure 3.3. Skewness is based on the the third standardized moment $E\{x^3\}$, which is a useful measure of the asymmety of the probability distribution function. Kurtosis is defined by the fourth standardized moment $E\{x^4\}$ that measures the heaviness of the tail of the distribution.

- To measure skewness in a distribution, we can use

$$g(s) = \frac{E\{(s-\mu)^3\}}{\sigma^3} \tag{3.9}$$

  where s is a random variable drawn from the distribution with mean μ and variance σ. To identify the third moment of the projection, we need to remove the first and second moments by sphering the data set, i.e.

$E\{x\} = 0, E\{(x - E\{x\})^2\} = 1$. Therefore we have $g(s) = s^3$ as an instantaneous measure of the skewness of the distribution.

- To measure kurtosis in a distribution, we can use

$$g(s) = \frac{E\{(s-\mu)^4\}}{\sigma^4} \tag{3.10}$$

and for sphered data set, we have $g(s) = s^4$ as above.

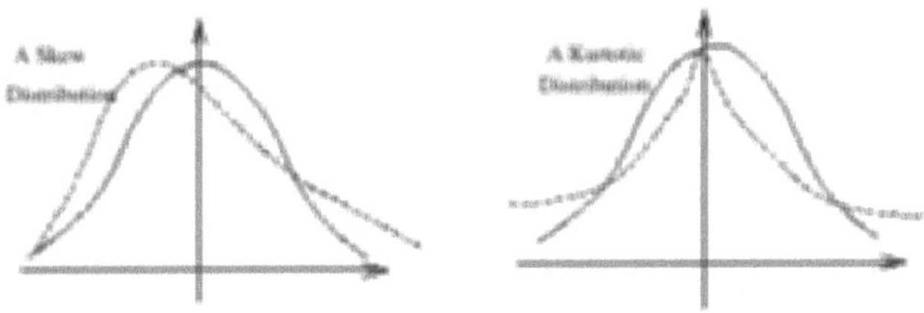

Fig. 3.3. Deviations from Gaussian Distribution represented by solid line. Left: negative skewness distribution. Right: positive kurtosis distribution.

Thus EPP can be thought of as an extension of the PCA method since it is a technique for finding projections that maximize some statistical feature of the data space. Apart from measuring high-order moments of the projection described above, Fyfe [91, 97] has implemented EPP using an artificial neural network based on Hebbian learning, where the only difference from the PCA network described in the previous subsection is that a function of output activations is calculated in (3.13). We have

$$y_i = \sum_{j=1}^{D} w_{ij} x_j \tag{3.11}$$

$$e_j = x_j - \sum_{i=1}^{M} w_{ij} y_i \tag{3.12}$$

$$r_i = f\left(\sum_{j=1}^{D} w_{ij} x_j\right) = f(y_i). \tag{3.13}$$

The update rule of EPP network is

$$\begin{aligned} \Delta w_{ij} &= \eta r_i e_j \\ &= \eta_t f\left(\sum_{k=1}^{D} w_{ik} x_k\right)\left\{x_j - \sum_{l=1}^{M} w_{lj} \sum_{p=1}^{D} w_{lp} x_p\right\}, \end{aligned} \tag{3.14}$$

which may be written in matrix form as

$$\Delta \mathbf{W} = \eta[\mathbf{I} - \mathbf{W}\mathbf{W}^T]\mathbf{x}f(\mathbf{x}^T\mathbf{W}) \qquad (3.15)$$

where $\mathbf{I}$ is the identity matrix and $f(.)$ is determined by the function $g(.)$ to be maximized.

3.1.3 Independent Component Analysis

Independent Component Analysis (ICA) has become, in recent years, a well established data analysis technique for data mining (see [122, 4, 102]). ICA defines a generative model for the observed multivariate data, which is typically given as a large database of samples. In the model, the data variables are assumed to be linear mixtures of some unknown latent variables, and the mixing system is also unknown. The latent variables are assumed non-Gaussian and mutually independent, and they are called the independent components of the observed data. These independent components (ICs), also called sources, can be found by ICA.

In detail, we consider ICA as the problem of transforming a set of D-dimensional observations $\mathbf{x}_1, \ldots, \mathbf{x}_N$ that are the result of a linear mixing of statistically independent sources $\mathbf{s}_1, \ldots, \mathbf{s}_N$ by $\mathbf{x} = \mathbf{A}\mathbf{s}$, into several components that are statistically independent by $\mathbf{y} = \mathbf{W}\mathbf{x}$. The independent sources $\mathbf{s}_j$ are often defined as latent variables, which means that they cannot be observed directly. At the same time, the mixing matrix $\mathbf{A}$ is also assumed to be unknown. Thus all we observe are the random variables $\mathbf{x}_i$, and we must estimate both the independent components $\mathbf{s}_i$ and the mixing matrix. It should be noted that there is always residual uncertainty in that each y_i cannot be guaranteed to correspond to the s_i i.e. the order of the independent components is not guaranteed and also the magnitude of the y_i is not guaranteed since if we multiply a signal by any scalar and divide the corresponding column of A by that scalar we have the same mixture as before.

Independent component analysis is originally motivated to deal with problems that are related to so-called *blind source separation* (BSS). Like the speaker in the cocktail party problem, an independent component can be used to represent the source from one speaker, and we know very little about the mixing matrix corresponding to the blending environment. The ICA method can be used to perform blind source separation.

ICA can also be seen as an extension to principal component analysis. Independent components are assumed statistically independent, while PCA aims to find a subpace in which the principal components are uncorrelated, which is a weaker form of independence. When performing ICA, *whitening* is frequently used as a pre-processing step in ICA to give the ICs up to an orthogonal transformation. Consequently, before performing an ICA algorithm, we firstly linearly transform the mixed data set $\mathbf{x}$ by multiplying it with some matrix $\mathbf{V}$,

$$\mathbf{z} = \mathbf{V}\mathbf{x} \qquad (3.16)$$

so the new vector is $\mathbf{z}$ whose components are uncorrelated and whose variances equal unity. Thus the covariance matrix of $\mathbf{z}$ equals the identity matrix, $E\{\mathbf{z}\mathbf{z}^T\} = I$. Although uncorrelatedness is weaker than independence and prewhitening only finds the ICs up to an orthogonal transformation, it is still helpful in that we can search for the demixing matrix W in the space of orthogonal matrices. A common way of whitening is to use the eigen-value decomposition of the covariance matrix $E\{\mathbf{x}\mathbf{x}^T\} = \mathbf{E}\mathbf{D}\mathbf{E}^T$, where $\mathbf{E}$ contains the eigenvectors of $E\{\mathbf{x}\mathbf{x}^T\}$ and $\mathbf{D}$ is a diagonal matrix of the eigenvalues, $\mathbf{D} = diag\{d_1, \ldots, d_D\}$. Then, we define the whitening matrix in the form of

$$\mathbf{V} = (\mathbf{D}^{-1/2})\mathbf{E}^T. \tag{3.17}$$

In the following subsection, we will review some methods of performing independent component analysis in detail.

ICA by maximizing Nongaussianity

Oja and Karhunen [178] have pointed out that it is usually difficult to measure the independence condition exactly in practice because the probability densities involved are unknown. Some ICA algorithms use approximate contrast functions (see Comon [51]). Moreau and Macchi [171] derived a relatively simple contrast function based on the idea that all independent components are as non-Gaussian as possible.

According to the *central limit theorem*, mixed observations become more Gaussian than any of the independent sources, so we can just measure the distribution of $\mathbf{Wx}$ and find an demixing matrix, $\mathbf{W}$, that maximizes the non-Gaussianity of $\mathbf{Wx}$. A popular way to measure the nongaussianity of a vector is using kurtosis $\texttt{kurt}(y)$, which is the fourth-order cumulant of a random variable y given by

$$\texttt{kurt}(y) = E\{y^4\} - 3(E\{y^2\})^2. \tag{3.18}$$

When y is normalized so that its variance is equal to one, $E\{y^2\} = 1$, the kurtosis is simplified to $E\{y^4\}-3$. For a Gaussian random variable y, the kurtosis is zero and random variables with positive kurtosis are super-Gaussian, and those with negative kurtosis are sub-Gaussian. The ICA algorithm is used to make the kurtosis of the final components as non-zero as possible, which gives components which are as non-Gaussian as possible. Thus for a whitened data vector, $\mathbf{z}$, the contrast function can be defined as

$$J_{kurt}(\mathbf{y}) = \sum_{i=1}^{N} |\texttt{cum}(y(i)^4)| = \sum_{i=1}^{N} |E\{y(i)^4\} - 3(E\{y(i)^2\})^2| \tag{3.19}$$

where $\mathbf{y} = \mathbf{Wz}$. Hyvarinen derives the following *fast-ICA* algorithm [122]:

$$\Delta\mathbf{w} \propto \texttt{sign}(\texttt{kurt}(\mathbf{w}^T\mathbf{z}))E\{\mathbf{z}(\mathbf{w}^T\mathbf{z})^3\} \tag{3.20}$$

$$\mathbf{w} \leftarrow \mathbf{w}/||\mathbf{w}|| \tag{3.21}$$

ICA by approximating negentropy

However Hyvarinen et al. [122] noted that kurtosis is not a robust method of measuring nongaussianity in some situations, because the value of kurtosis may depend on only a few observations in the tails of the distribution. An alternative way is by approximating negentropy. Negentropy is a concept from information theory. The basic idea is that since a Gaussian variable has the largest entropy among all random variables of equal variance, we can use entropy to measure non-Gaussianity.

According to [122], negentropy J is defined as

$$J(\mathbf{y}) = H(\mathbf{y}_{gauss}) - H(\mathbf{y}). \tag{3.22}$$

where $\mathbf{y}_{gauss}$ is a gaussian random variable with the same variance as $\mathbf{y}$. A possible way to approximate negentropy is

$$J(y) \propto [E\{G(y)\} - E\{G(\nu)\}]^2 \tag{3.23}$$

where $G(y)$ is a nonquadratic function that does not grow fast and ν is a standardized Gaussian random variable. Examples of $G(y)$ are

$$G_1(y) = \frac{1}{a_1} \log\cosh a_1 y \tag{3.24}$$

$$G_2(y) = -\exp(-y^2/2) \tag{3.25}$$

$$G_3(y) = \tanh(a_1 y) \tag{3.26}$$

$$G_4(y) = y\exp(-y^2/2) \tag{3.27}$$

$$G_5(y) = y^3 \tag{3.28}$$

where $a_1 \in [1, 2]$ and is often taken as $a_1 = 1$.

Thus the negentropy can be maximized by taking the gradient of the approximation of negentropy in (3.23) with respect to $\mathbf{w}$, and for prewhitened data, the algorithm [122] can be followed

$$\Delta\mathbf{w} \propto \gamma E\{\mathbf{z}G(\mathbf{w}^T\mathbf{z})\} \tag{3.29}$$

$$\mathbf{w} \leftarrow \mathbf{w}/||\mathbf{w}|| \tag{3.30}$$

where $\gamma = E\{G(\mathbf{w}^T\mathbf{z})\} - E\{G(\nu)\}$.

ICA by Maximum Likelihood Estimation

Maximum Likelihood Estimation (MLE) has been widely studied to estimate independent component analysis (see [122, 121, 120]). Based on the density of a linear transform $\mathbf{x} = \mathbf{A}\mathbf{s}$, given by

$$p_{\mathbf{x}}(\mathbf{x}) = \frac{1}{|\det \mathbf{A}|}\mathbf{p}_{\mathbf{s}}(\mathbf{A}^{-1}\mathbf{x}), \tag{3.31}$$

the density $p_{\mathbf{x}}$ of the mixed vector can be formulated as

$$p_x(\mathbf{x}) = |\det \mathbf{W}| p_s(\mathbf{s}) = |\det \mathbf{W}| \prod_i p_i(s_i) \tag{3.32}$$

where $\mathbf{W} = \mathbf{A}^{-1}$. Then assuming there are N observations of $\mathbf{x}$, we obtain the log-likelihood as

$$\log L(\mathbf{W}) = \sum_{t=1}^{N} \sum_{i=1}^{D} \log p_i(\mathbf{w}_i^T \mathbf{x}(t)) + N \log |\det \mathbf{W}| \tag{3.33}$$

Bell and Sejnowski [22] have derived the stochastic gradient of the log-likelihood, known as the Bell-Sejnowski algorithm

$$\Delta \mathbf{W} \propto \frac{1}{N} \frac{\partial \log L}{\partial \mathbf{W}} = [\mathbf{W}^T]^{-1} + g(\mathbf{W}\mathbf{x})\mathbf{x}^T, \tag{3.34}$$

where $\mathbf{g}(\mathbf{W}\mathbf{x}) = (g_1(\mathbf{w}_1\mathbf{x}), \dots, g_D(\mathbf{w}_n\mathbf{x}))$ consists of the score function g_i of the distribution of $\mathbf{s}_i$, $g_i = (\log p_i)' = \frac{p_i'}{p_i}$.

ICA by Non-linear PCA

It [177] has been proved that for *prewhitened data*, there is an exact relationship between non-linear PCA and ICA. For non-linear PCA, one criterion is the *so-called* "least mean-square error reconstruction" (LMSER) first considered by Xu [235] in the form of

$$J(\mathbf{w}_1, \dots, \mathbf{w}_D) = J(\mathbf{W}) = E\{||\mathbf{z} - \mathbf{W}^T \mathbf{g}(\mathbf{W}\mathbf{z})||^2\}. \tag{3.35}$$

We have [177]:

$$\begin{aligned}
||\mathbf{z} - \mathbf{W}^T \mathbf{g}(\mathbf{W}\mathbf{z})||^2 &= [\mathbf{z} - \mathbf{W}^T \mathbf{g}(\mathbf{W}\mathbf{z})]^T [\mathbf{z} - \mathbf{W}^T \mathbf{g}(\mathbf{W}\mathbf{z})] && (3.36)\\
&= [\mathbf{z} - \mathbf{W}^T \mathbf{g}(\mathbf{W}\mathbf{z})]^T \mathbf{W}^T \mathbf{W} [\mathbf{z} - \mathbf{W}^T \mathbf{g}(\mathbf{W}\mathbf{z})] && (3.37)\\
&= ||\mathbf{W}\mathbf{z} - \mathbf{W}\mathbf{W}^T \mathbf{g}(\mathbf{W}\mathbf{z})||^2 && (3.38)\\
&= ||\mathbf{y} - \mathbf{g}(\mathbf{y})||^2 && (3.39)\\
&= \sum_{i=1}^{N} [y(i) - g_i(y(i))]^2. && (3.40)
\end{aligned}$$

with $\mathbf{y} = \mathbf{W}\mathbf{z}$. Thus the criterion (3.35) becomes

$$J(\mathbf{W}) = \sum_{i=1}^{N} E\{[y_i - g_i(y(i))]^2\}, \tag{3.41}$$

and by defining an odd quadratic function on $g_i(y)$,

$$g_i(y(i)) = \begin{cases} y^2(i) + y(i), & y(i) \geq 0 \\ -y^2(i) + y(i), & y(i) < 0 \end{cases}, \tag{3.42}$$

the criterion (3.41) becomes

$$J(\mathbf{W}) = \sum_{i=1}^{N} E\{y^4(i)\}. \tag{3.43}$$

Under the condition that the data has been whitened, the variance $E\{y^2(i)\} = 1$, the second term of (3.19) can be dropped, which is exactly equivalent to (3.43). Then the problem of ICA can be turned to a special form of the non-linear PCA problem.

3.1.4 Canonical Correlation Analysis

Canonical Correlation Analysis (CCA) is a statistical technique used when there are two sets of data which we believe have some underlying correlation. Different from principal component analysis and exploratory projection analysis that aim to investigate the intra-relationships within one data set, CCA is used to deal with the relationship between two sets of variables.

In classical CCA, we have two sets of input data, $\mathbf{x}_1$ and $\mathbf{x}_2$. We attempt to find a linear combination of the variables which give us maximum correlation between the combinations. Thus, let

$$y_1 = \mathbf{w}_1\mathbf{x}_1 = \sum_j w_{1j}x_{1j} \tag{3.44}$$

$$y_2 = \mathbf{w}_2\mathbf{x}_2 = \sum_j w_{2j}x_{2j} \tag{3.45}$$

where x_{ij} represents the jth element of $\mathbf{x}_i$. We wish to find values of $\mathbf{w_1}$ and $\mathbf{w_2}$ that maximise the correlation between y_1 and y_2. The standard statistical method [159] lies in defining

$$\Sigma_{11} = E\{(\mathbf{x}_1 - \mu_1)(\mathbf{x}_1 - \mu_1)^T\} \tag{3.46}$$

$$\Sigma_{22} = E\{(\mathbf{x}_2 - \mu_2)(\mathbf{x}_2 - \mu_2)^T\} \tag{3.47}$$

$$\Sigma_{12} = E\{(\mathbf{x}_1 - \mu_1)(\mathbf{x}_2 - \mu_2)^T\} \tag{3.48}$$

and

$$K = \Sigma_{11}^{-\frac{1}{2}}\Sigma_{12}\Sigma_{22}^{-\frac{1}{2}}. \tag{3.49}$$

Then we perform a Singular Value Decomposition of K, and we have

$$K = (\alpha_1, \ldots, \alpha_k)D(\beta_1, \ldots, \beta_k)^T \tag{3.50}$$

where α_i and β_i are the eigenvectors of KK^T and K^TK respectively and D is the diagonal matrix of eigenvalues. The ith canonical correlation vectors are given by

$$\mathbf{w}_{1i} = \Sigma_{11}^{-\frac{1}{2}}\alpha_i \tag{3.51}$$

$$\mathbf{w}_{2i} = \Sigma_{22}^{-\frac{1}{2}}\beta_i \tag{3.52}$$

Lai and Fyfe [144, 142, 143] have implemented canonical correlation analysis with a neural network by maximizing the correlation $E(y_1y_2)$ where $E()$ is the expected value which will be taken over the joint distribution of $\mathbf{x_1}$ and $\mathbf{x_2}$. The CCA problem is thus regarded as the problem of maximising the function $g_1(\mathbf{w}_1|\mathbf{w}_2) = E(y_1y_2)$ as a function of the weights $\mathbf{w_1}$ and $g_2(\mathbf{w}_2|\mathbf{w}_1)$ for the weights $\mathbf{w_2}$, respectively. The optimization function is defined by,

$$J = E(y_1y_2) + \frac{1}{2}\lambda_1(y_1^2 - 1) + \frac{1}{2}\lambda_2(y_2^2 - 1) \tag{3.53}$$

which would be optimized by changing the weights $\mathbf{w_1}$ and $\mathbf{w_2}$ separately with gradient ascent in the rule of

$$\Delta w_{1j} = \eta x_{1j}(y_2 - \lambda_1 y_1) \tag{3.54}$$

$$\Delta\lambda_1 = \eta_0(y_1^2 - 1) \tag{3.55}$$

for $\mathbf{w_1}$ and similarly with $\mathbf{w_2}$ and λ_2.

3.1.5 Deflationary Orthogonalization methods

In general, the projection methods described in the previous subsections are sometimes called *one-unit* algorithms, by which we mean that we only estimate one component. However, it is sometimes necessary to identify more than one component. The key to extending the projection methods to estimate more components is based on the fact that the weight vectors $\mathbf{w}_i$ corresponding to different components are orthogonal in the projection subspace. The two most common orthogonalization algorithms are the Gram-Schmidt method [245] and symmetric orthogonalization [77]. In this book, we have more interest in the former.

Deflationary orthogonalization by the Gram-Schmidt method [245] is a simple and popular way to orthogonalize the weight vectors $\mathbf{w}_1, \ldots, \mathbf{w}_N$. Provided that we have estimated $j-1$ weight vectors corresponding to the first $j-1$ components and we run the same algorithm for the next weight vector $\mathbf{w}_j$, we estimate $\mathbf{w}_j$ first and subtract the projection $(\mathbf{w}_j^T\mathbf{w}_k)\mathbf{w}_k, k = 1, \ldots, j-1$ from $\mathbf{w}_j$, so we have a new weight vector $\mathbf{w}_j^{'}$ and finally $\mathbf{w}_j$ is set to be $\mathbf{w}_j^{'}$. More precisely, we follow the steps to estimate $\mathbf{w}_j$:

1. Randomly initialize the weight vector $\mathbf{w}_j$.
2. Perform the same algorithm on $\mathbf{w}_j$ as performed on the previous $j-1$ components.

3. Do the following orthogonalization:

$$\mathbf{w}_j \leftarrow \mathbf{w}_j - \sum_{k=1}^{j-1} (\mathbf{w}_j^T \mathbf{w}_k)\mathbf{w}_k \tag{3.56}$$

4. Normalize $\mathbf{w}_j$ by $\mathbf{w}_j \leftarrow \mathbf{w}_j/||\mathbf{w}_j||$.
5. Set $j = j + 1$ and go back to the first step.

3.2 Kernel Methods

Kernel methods [52] are a class of algorithms for pattern analysis, whose best known member is the Support Vector Machine (SVM). The basic idea of kernel methods is to map the data set first into a high-dimensional feature space in which a variety of linear methods can be used to find relations in the data. Specifically the input vectors $\mathbf{x} \in \Re^D$ from an input space are mapped into some feature space $\mathcal{F}$ by $\Phi : \Re^D \rightarrow \mathcal{F}$ through a kernel function $k : \Re^D \times \Re^D \rightarrow \Re$, where the scalar product can be easily calculated with the kernel function (we often call it the kernel tick) $k(\mathbf{x}_i, \mathbf{x}_j) = \Phi^T(\mathbf{x}_i)\Phi(\mathbf{x}_j)$. Some common kernels include

- Polynomial (homogeneous): $k(\mathbf{x}_i, \mathbf{x}_j) = (\mathbf{x}_i \cdot \mathbf{x}_j)^d$
- Polynomial (heterogeneous): $k(\mathbf{x}_i, \mathbf{x}_j) = (\mathbf{x}_i \cdot \mathbf{x}_j + 1)^d$
- Radial Basis Function: $k(\mathbf{x}_i, \mathbf{x}_j) = \exp(-\gamma||\mathbf{x}_i - \mathbf{x}_j||^2)$, for $\gamma > 0$
- Sigmoid: $k(\mathbf{x}_i, \mathbf{x}_j) = \tanh(\kappa \mathbf{x}_i \cdot \mathbf{x}_j + c)$, for some $\kappa > 0$ and $c < 0$

The use of kernel functions enables us to operate in the feature space without ever computing the coordinates of the data in that space, but rather by simply computing the inner products between the images of all pairs of data in the feature space. This operation is often computationally cheaper than the explicit computation of the coordinates and may even be done when it is not possible to calculate these coordinates. Moreover, any linear operation in the feature space corresponds to a non-linear operation in the original data space.

Kernel methods have been widely used to derive various methods originating in multivariate statistics. Prominent examples are kernel principal component analysis [198], kernel discriminant analysis [167] and variants of chemometric regression methods like kernel ridge regression and kernel partial least squares regression [192], and many others (see [39, 119, 201]). Furthermore, several authors have studied the construction of a kernel variant of CCA and proposed quite different algorithms [145, 163, 223, 10]. In the following sections, we review two projection models based on the kernel method: Kernel Principal Component Analysis and Kernel Canonical Correlation Analysis.

3.2.1 Kernel Principal Component Analysis

Scholkopf et al. [170, 191, 168, 169, 198, 199, 207, 98] have shown that Principal Component Analysis can be regarded as a kernel-based problem, which

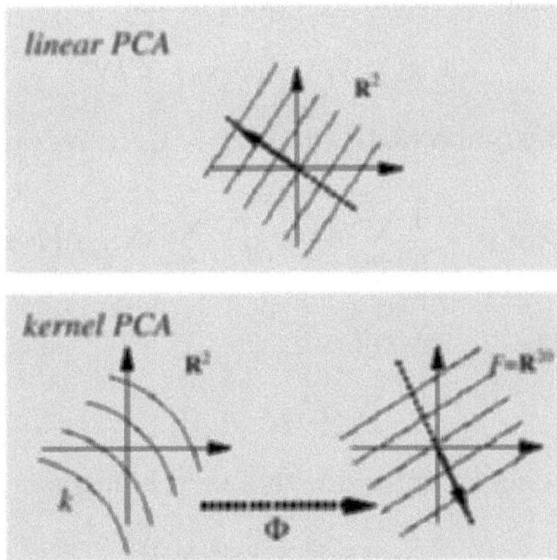

Fig. 3.4. The basic idea of Kernel PCA. Kernel PCA performs linear PCA (top) in the high-dimensional feature space $\mathcal{F}$ (bottom right), which is nonlinearly related to the data space. The contour lines of constant projections in the feature space become nonlinear in the data space (bottom left)[198].

is *so-called* Kernel Principal Component Analysis (KPCA). This subsection mainly follows the analysis in [198].

In Subsection 3.1.1, we presented the PCA method which attempts to find the directions of maximal variance in data space. Based on the idea of kernel methods, such computation as shown in Figure 3.4, is performed in a feature space which is related to the data space through a non-linear mapping

$$\Phi : \Re^D \to \mathcal{F}. \tag{3.57}$$

We assume the data in feature space is centered, i.e. $\sum_{k=1}^{N} \Phi(\mathbf{x}_k) = 0$. By defining the covariance matrix in $\mathcal{F}$,

$$\bar{\mathbf{C}} = \frac{1}{N}\sum_{j=1}^{N} \Phi(\mathbf{x}_j)\Phi(\mathbf{x}_j)^T \tag{3.58}$$

we can maximize the variance in feature space by finding eigenvalues $\bar{\lambda} \geq 0$ and eigenvectors $\mathbf{W} \in \mathcal{F}$ satisfying

$$\bar{\mathbf{C}}\mathbf{W} = \bar{\lambda}\mathbf{W} \tag{3.59}$$

where $\mathbf{W}$ lies in the span of $\Phi(\mathbf{x}_1), \ldots, \Phi(\mathbf{x}_N)$ and each direction $\mathbf{w}$ of $\mathbf{W}$ is parameterized by $N \times 1$ coefficient vector $\alpha = (\alpha_1, \ldots, \alpha_N)^T$ such that

$$\mathbf{w} = \sum_{i=1}^{N} \alpha_i \Phi(\mathbf{x}_i). \tag{3.60}$$

Multiplying both sides of (3.59) by $\Phi(\mathbf{x}_k)$, we have

$$\bar{\lambda}(\Phi(\mathbf{x}_k) \cdot \mathbf{W}) = (\Phi(\mathbf{x}_k) \cdot \bar{\mathbf{C}}\mathbf{W}). \tag{3.61}$$

By substituting (3.60) into (3.61), for all $k = 1, \ldots, N$, we get

$$\bar{\lambda} \sum_{i=1}^{N} \alpha_i (\Phi(\mathbf{x}_k) \cdot \Phi(\mathbf{x}_i)) = \frac{1}{N} \sum_{i=1}^{N} \alpha_i (\Phi(\mathbf{x}_k) \cdot \sum_{j=1}^{N} \Phi(\mathbf{x}_j))(\Phi(\mathbf{x}_j) \cdot \Phi(\mathbf{x}_i)). \tag{3.62}$$

We can rewrite (3.62) in form of

$$N\bar{\lambda}K\alpha = K^2\alpha \tag{3.63}$$

where the dot product is replaced with a kernel function,

$$K_{ij} = (\Phi(\mathbf{x}_i) \cdot \Phi(\mathbf{x}_k)). \tag{3.64}$$

Scholkopf [198] has shown that the solutions of this equation also satisfy

$$N\bar{\lambda}\alpha = K\alpha \tag{3.65}$$

so that α is therefore an eigenvector of K. So we get eigenvectors of the kernel matrix and use this in (3.60) to get $\mathbf{w}$, the eigenvectors of the covariance matrix (3.59).

For Kernel PCA, we must normalize the eigenvectors in feature space, $\mathbf{W}$, to be of unit length. According to (3.60), $\mathbf{w}_i$ is an eigenvector of $\bar{C}$ that lies in the feature space $\mathcal{F}$ and is spanned by $\Phi(\mathbf{x}_1), \ldots, \Phi(\mathbf{x}_M)$. Therefore,

$$\mathbf{w}_i^T \mathbf{w}_i = \sum_{j,k=1}^{N} \alpha_j^i \Phi(\mathbf{x}_j)^T \Phi(\mathbf{x}_k) \alpha_k^i \tag{3.66}$$

$$= \sum_{j,k=1}^{N} \alpha_j^i K_{jk} \alpha_k^i \tag{3.67}$$

$$= \alpha^i \cdot (K\alpha^i) \tag{3.68}$$

$$= \lambda_i (\alpha^i \cdot \alpha^i) \tag{3.69}$$

To normalize the eigenvectors in the feature space, i.e. $\mathbf{w}_i^T \mathbf{w}_i = 1$, we can normalize the eigenvectors of K, α^i, by dividing each by the square root of their corresponding eigenvalues.

Then given any data point $\mathbf{x}$, we can extract its principal components in feature space, $\mathcal{F}$, by

$$\mathbf{w}_i \cdot \Phi(\mathbf{x}) = \sum_{j=1}^{N} \alpha_j^i \Phi(\mathbf{x}_j)^T \Phi(\mathbf{x}) = \sum_{j=1}^{N} \alpha_j^i K(\mathbf{x}_j, \mathbf{x}). \tag{3.70}$$

3.2.2 Kernel Canonical Correlation Analysis

In Subsection 3.1.4, we presented the essential idea of canonical correlation analysis. CCA finds a linear transformation of a pair of multi-variates such that the correlation between the sets of data is maximized. From an information theoretical point of view, with a Gaussian distribution, the transformation maximizes the mutual information between extracted features. However, if there is a nonlinear relation between the variates, linear CCA does not always extract useful features. In the previous section, we can see that any operation which can be defined in terms of dot products can be kernelized, which is used for Kernel Canonical Correlation Analysis (KCCA) (see [145, 163, 223]).

In this subsection, we review the Kernel CCA method based on the study in [145], in which kernel CCA is performed in the feature space formed from a nonlinear mapping from the data space. Thus, consider mapping the data to a high dimensional feature space, $\mathcal{F}$. We have

$$\Sigma_{11} = E\{(\Phi(\mathbf{x}_1) - \mu_1)(\Phi(\mathbf{x}_1) - \mu_1)^T\} \tag{3.71}$$

$$\Sigma_{22} = E\{(\Phi(\mathbf{x}_2) - \mu_2)(\Phi(\mathbf{x}_2) - \mu_2)^T\} \tag{3.72}$$

$$\Sigma_{12} = E\{(\Phi(\mathbf{x}_1) - \mu_1)(\Phi(\mathbf{x}_2) - \mu_2)^T\} \tag{3.73}$$

where $\mu_i = E(\Phi(\mathbf{x}_i))$. We then centre the data in feature space according to [198] and define

$$\Sigma_{11} = E\{\Phi(\mathbf{x}_1)\Phi(\mathbf{x}_1)^T\} \tag{3.74}$$

$$\Sigma_{22} = E\{\Phi(\mathbf{x}_2)\Phi(\mathbf{x}_2)^T\} \tag{3.75}$$

$$\Sigma_{12} = E\{\Phi(\mathbf{x}_1)\Phi(\mathbf{x}_2)^T\}. \tag{3.76}$$

We wish to find those values, $\mathbf{w}_1$ and $\mathbf{w}_2$, which would maximize

$$\mathbf{w}_1^T \Sigma_{12} \mathbf{w}_2 \tag{3.77}$$

in terms of the constraints $\mathbf{w}_1^T \Sigma_{11} \mathbf{w}_1 = 1$ and $\mathbf{w}_2^T \Sigma_{22} \mathbf{w}_2 = 1$. In practice, Σ_{12} is approximated by the sample average value $\frac{1}{N}\sum_i \Phi(\mathbf{x}_{1i})\Phi(\mathbf{x}_{2i})$. The $\mathbf{w}_1$ and $\mathbf{w}_2$ are in feature space and spanned by $\Phi(\mathbf{x}_{11}), \ldots, \Phi(\mathbf{x}_{1N}), \Phi(\mathbf{x}_{21}), \ldots, \Phi(\mathbf{x}_{2N})$ and can be expressed by

$$\mathbf{w}_1 = \sum_{i=1}^{N} \alpha_{1i}\Phi(\mathbf{x}_{1i}) + \sum_{i=1}^{N} \alpha_{2i}\Phi(\mathbf{x}_{2i}), \tag{3.78}$$

$$\mathbf{w}_2 = \sum_{i=1}^{N} \beta_{1i}\Phi(\mathbf{x}_{1i}) + \sum_{i=1}^{N} \beta_{2i}\Phi(\mathbf{x}_{2i}). \tag{3.79}$$

Substituting (3.78) and (3.79) into the criteria (3.77), we obtain

$$\mathbf{w}_1^T \Sigma_{12} \mathbf{w}_2 = \frac{1}{N} \sum_{k,i} \alpha_k \cdot \Phi^T(\mathbf{x}_k) \Phi(\mathbf{x}_{1i}) \sum_l \beta_l \cdot \Phi^T(\mathbf{x}_{2i}) \Phi(\mathbf{x}_l) \tag{3.80}$$

and with the constraints

$$\mathbf{w}_1^T \Sigma_{11} \mathbf{w}_1 = \frac{1}{N} \sum_{k,i} \alpha_k \cdot \Phi^T(\mathbf{x}_k) \Phi(\mathbf{x}_{1i}) \sum_l \alpha_l \cdot \Phi^T(\mathbf{x}_{1i}) \Phi(\mathbf{x}_l) = 1 \tag{3.81}$$

$$\mathbf{w}_2^T \Sigma_{22} \mathbf{w}_2 = \frac{1}{N} \sum_{k,i} \beta_k \cdot \Phi^T(\mathbf{x}_k) \Phi(\mathbf{x}_{2i}) \sum_l \beta_l \cdot \Phi^T(\mathbf{x}_{2i}) \Phi(\mathbf{x}_l) = 1. \tag{3.82}$$

Again, using $(K_1)_{ij} = \Phi^T(\mathbf{x}_{1j})\Phi(\mathbf{x}_{1i})$ and $(K_2)_{ij} = \Phi^T(\mathbf{x}_{2j})\Phi(\mathbf{x}_{2i})$, the criteria we wish to optimize becomes $\alpha^T K_1 K_2^T \beta$ subject to the constraints $\alpha^T K_1 K_1^T \alpha = 1$ and $\beta^T K_2 K_2^T \beta = 1$. We define $\Gamma_{11} = K_1 K_1^T$, $\Gamma_{22} = K_2 K_2^T$ and $\Gamma_{12} = K_1 K_2^T$, and we solve the problem by forming the matrix $K = \Gamma_{11}^{-\frac{1}{2}} \Gamma_{12} \Gamma_{22}^{-\frac{1}{2}}$ and perform a singular value decomposition on it. We have

$$K = (\gamma_1, \ldots, \gamma_k) D(\theta_1, \ldots, \theta_k)^T \tag{3.83}$$

where γ_i and θ_i are the standardized eigenvectors of KK^T and $K^T K$ respectively and D is the diagonal matrix of eigenvalues.

Then the first canonical correlation vectors in feature space is given by

$$\alpha_1 = \Gamma_{11}^{-\frac{1}{2}} \gamma_1 \tag{3.84}$$

$$\beta_1 = \Gamma_{22}^{-\frac{1}{2}} \theta_1 \tag{3.85}$$

and subsequent canonical correlation vectors are defined in terms of the subsequent eigenvectors, γ_i and θ_i.

For any new value $\mathbf{x}_1$, we may calculate its canonical correlation components in the feature space by

$$\mathbf{w}_1 \cdot \Phi(\mathbf{x}_1) = \sum_i \alpha_i \Phi(\mathbf{x}_i) \Phi(\mathbf{x}_1) = \sum_i \alpha_i K_1(\mathbf{x}_i, \mathbf{x}_1) \tag{3.86}$$

3.3 Latent Variable Models

Latent variable modeling [33] is a powerful approach where a set of observed data variables are associated with additional latent variables in some lower dimensional embedded space. The corresponding distribution of the observed variables can be formed through a joint distribution over the data and latent variables. Based on latent variable models, [219] and [146] provide a *probabilistic* formulation of the well-known technique of principal component analysis (PCA). Probabilistic canonical correlation analysis (CCA) has been derived by [94]. The latent variable model has also been generalized to nonlinear topology-preserving latent variable models in [37, 96, 133].

3.3.1 Density Modeling and Latent Variables

One of the most popular approaches to the problem of density estimation is to apply parametric models where a specific form for the density containing a number of adaptive parameters is proposed. Then values of these parameters are determined by the data set $D = \{\mathbf{t}_1, \ldots, \mathbf{t}_N\}$. The most widely used parametric model is the Gaussian model given by

$$p(\mathbf{t}|\mu, \Sigma) = (2\pi)^{-q/2}|\Sigma|^{-1/2}\exp\{-\frac{1}{2}(\mathbf{t}-\mu)\Sigma^{-1}(\mathbf{t}-\mu)^T\} \tag{3.87}$$

where μ is the mean, Σ is the covariance matrix, and $|\Sigma|$ is the determinant of Σ. Maximum likelihood estimation can be used to set the values of these parameters by considering the log-probability of the observed data given the parameters. For a Gaussian model, we have the log-likelihood function in the form of

$$L(\mu, \Sigma) = \ln p(D|\mu, \Sigma) = \sum_{n=1}^{N} \ln p(\mathbf{t}_n|\mu, \Sigma) \tag{3.88}$$

where $p(D|\mu, \Sigma)$ is called the *likelihood* function, which is viewed as a function of μ and Σ. Therefore maximization of the log-likelihood with respect to the parameters leads to the set of the values of the parameters that are most likely to match the data set. For example, for the Gaussian distribution, the maximum likelihood solutions are given by

$$\hat{\mu} = \frac{1}{N}\sum_{n=1}^{N} \mathbf{t}_n \tag{3.89}$$

$$\hat{\Sigma} = \frac{1}{N}\sum_{n=1}^{N} (\mathbf{t}_n - \hat{\mu})(\mathbf{t}_n - \hat{\mu})^T \tag{3.90}$$

However, considering the number of free parameters in (3.87), since Σ is symmetric, it contains $D(D+1)/2$ independent parameters and there are an additional D independent parameters in μ, so there are $D(D+3)/2$ parameters totally. For all D this number grows quadratically, and requires a large number of data points so as to ensure that the maximum likelihood solution is well estimated.

The latent variable model is a powerful approach to probabilistic modeling that involves supplementing the set of observed variables with additional latent variables. The goal of a latent variable model is to denote the distribution $p(\mathbf{t})$ of the variables $t_1, \ldots, t_D$ in terms of a smaller number of latent variables $\mathbf{x} = (x_1, \ldots, x_q)$ where $q < D$. Typically we specify a latent variable model relating a set of latent variables, $\mathbf{X} \in \Re^{N \times q}$, to a set of observed variables, $\mathbf{T} \in \Re^{N \times D}$, through a set of parameters. The joint distribution $p(\mathbf{t}, \mathbf{x})$ can be expressed by the product of the marginal distribution $p(\mathbf{x})$ of the latent variables and the conditional distribution $p(\mathbf{t}|\mathbf{x})$ of the data variables given the latent variables, which is

$$p(\mathbf{t}, \mathbf{x}) = p(\mathbf{x})p(\mathbf{t}|\mathbf{x}) \tag{3.91}$$

Denoting $\mathbf{y}(\mathbf{x}; \mathbf{w})$ as a function of the latent variable $\mathbf{x}$ with parameters $\mathbf{w}$, in terms of mapping from the latent variables to data variables, we have

$$\mathbf{t} = \mathrm{y}(\mathbf{x}; \mathbf{w}) + \epsilon \tag{3.92}$$

where $\mathbf{y}$ is a function of the latent variables $\mathbf{x}$ with parameters $\mathbf{w}$ and ϵ is a noise model whose components are uncorrelated. From a geometrical point of view, we regard the function $\mathbf{y}(\mathbf{x}; \mathbf{w})$ as defining a manifold embedded in data space given by the image of the latent space, as shown in Figure 3.5 [33]. Thus by specifying the distribution $p(\epsilon)$, the mapping $\mathbf{y}(\mathbf{x}; \mathbf{w})$ and the marginal distribution $p(\mathbf{x})$, the latent variable model for the distribution $p(\mathbf{t})$ of the data set is obtained by marginalizing over the latent variables

$$p(\mathbf{t}) = \int p(\mathbf{t}|\mathbf{x})p(\mathbf{x})d\mathbf{x} \tag{3.93}$$

The most common latent variable model is *factor analysis* [20], where the mapping is linear:

$$\mathbf{t} = \mathbf{W}\mathbf{x} + \mu + \epsilon \tag{3.94}$$

The $D \times q$ matrix W relates the latent variables and data variables and the parameter vector μ permits the model to have non-zero mean. Defining $\mathbf{x} \sim \mathcal{N}(0, I)$ so that the latent variables are independent and Gaussian distributed with unit variance, and the noise model is $\epsilon \sim \mathcal{N}(0, \psi)$, the distribution of the observations corresponds to the Gaussian distribution, $\mathbf{t} \sim \mathcal{N}(\mu, \mathbf{W}\mathbf{W}^T + \psi)$. The latent variables are then marginalized and the parameters are found through maximising the likelihood via an iterative procedure.

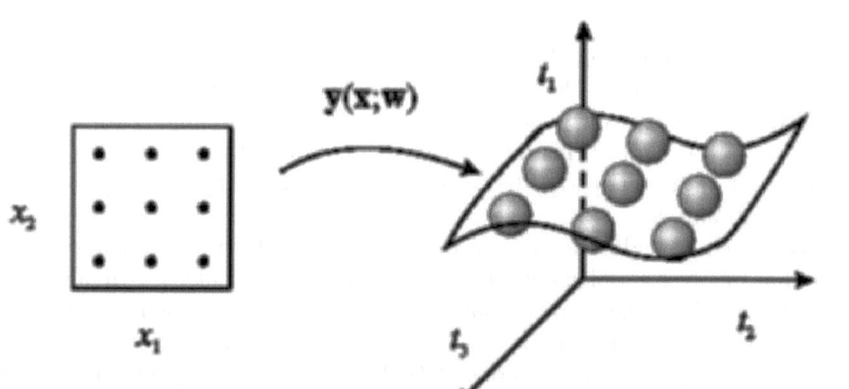

Fig. 3.5. The function $\mathbf{y}(\mathbf{x}; \mathbf{w})$ defines a manifold embedded in data space given by the image of the latent space [33]

3.3.2 Probabilistic Principal Component Analysis

We stated in Subsection 3.1.1 that PCA is a well-established statistical technique that has been widely studied. However, a general feature of these definitions of PCA is that they are not based on an associated probabilistic model for the observed data. A certain link [20] has been established between the standard factor analysis model and the PCA problem, where for PCA, the residual variances of the zero mean noise model, $\psi_i = \sigma^2$ for ϵ, are constrained to be equal, so that the subspace found by the maximum-likelihood estimation would correspond to the principal subspace of the observed data. Based on such a consideration, Young and Whittle [237, 229] derived the Young-Whittle factor analysis model under the condition that the residual variance σ^2 is presumed known. Anderson and Basilevsky [8, 21] show that the covariance $\mathbf{W}\mathbf{W}^T + \psi$ of the observations can be exactly calculated and both $\mathbf{W}$ and σ^2 may be determined analytically without resorting to iteration. However, Tipping and Bishop [219] pointed out that the presumption of an additive noise model or an exact estimation of the covariance model is generally undesirable. Tipping and Bishop [219] have derived a probabilistic version of PCA based on a Gaussian latent variable model in which the maximum likelihood estimation of the parameters, $\mathbf{W}_{\mathbf{ML}}$ and $\sigma^2_{\mathbf{ML}}$, is done by solving the eigenvalue problem on the covariance matrix of the data set and the solution corresponds to principal component analysis.

Consider a set of centered D-dimensional data $\mathbf{t} = [\mathbf{t}_1, \dots, \mathbf{t}_N]^T$. The latent variable model consists of the q-dimensional latent variables with the same linear mapping as (3.94). We take the noise variable ϵ in (3.94) to be an independent sample from a Gaussian distribution with zero mean and covariance, $\sigma^2 I$, $\epsilon_n \sim \mathcal{N}(0, \sigma^2 I)$, which implies that the $\mathbf{x}$-conditional probability distribution over the data space is given by

$$p(\mathbf{t}|\mathbf{x}) \sim \mathcal{N}(\mathbf{W}\mathbf{x} + \mu, \sigma^2\mathbf{I}) \tag{3.95}$$

We regard $p(\mathbf{x})$ as a *prior* distribution over the latent variables, which is defined by

$$p(\mathbf{x}) \sim \mathcal{N}(0, I) \tag{3.96}$$

Following (3.93), we then obtain the marginal distribution of $\mathbf{t}$ in the form

$$p(\mathbf{t}) \sim \mathcal{N}(\mu, \mathbf{C}) \tag{3.97}$$

where $\mathbf{C}$ is the covariance model and specified by $\mathbf{C} = \mathbf{W}\mathbf{W}^{\mathbf{T}} + \sigma^2\mathbf{I}$. The corresponding log-likelihood is then defined

$$\mathcal{L} = -\frac{N}{2}\{D\ln(2\pi) + \ln|\mathbf{C}| + \mathtt{tr}(\mathbf{C}^{-1}\mathbf{S})\} \tag{3.98}$$

where $\mathbf{S}$ is the covariance matrix of the data $\mathbf{t}_n$,

$$\mathbf{S} = \frac{1}{N}\sum_{n=1}^{N}(\mathbf{t}_n - \mu)(\mathbf{t}_n - \mu)^T. \tag{3.99}$$

Thus, the parameters to be estimated by the maximun likelihood estimation are $\mu, \mathbf{W}$ and σ^2. The estimator for μ is given by the mean of the data set and estimates for $\mathbf{W}$ and σ^2 can be performed by iterative maximization of $\mathcal{L}$. It [219] has been shown that the likelihood (3.98) is maximized when

$$\mathbf{W}_{\mathbf{ML}} = \mathbf{U}_q(\Lambda_q - \sigma^2\mathbf{I})^{1/2}\mathbf{R}, \tag{3.100}$$

where the components of the $D \times q$ matrix $\mathbf{U}_q$ are the principal eigenvectors of $\mathbf{S}$, the sample covariance matrix, with the corresponding eigenvalues $\lambda_1, \ldots, \lambda_q$ in the $q \times q$ diagonal matrix Λ_q and $\mathbf{R}$ is the rotation matrix. For $\mathbf{W} = \mathbf{W}_{\mathbf{ML}}$, the optimization of σ^2 is given by

$$\sigma^2_{\mathbf{ML}} = \frac{1}{D-q} \sum_{j=q+1}^{D} \lambda_j. \tag{3.101}$$

We can see the latent variable model defined by (3.94) therefore involves a mapping from the latent space into the *principal subspace* of the data set.

3.4 Conclusions

In this chapter, we have introduced several of the most common projection methods which have been used for different projection problems. We can not say which projection method is the best one over the data set because different projection methods have their own objectives and are applied to solve different projection problems. However, the main conclusion is that when trying to investigate the structure of the data set with a particular objective, we not only need to make clear the objective itself, but also we need to emphasize more defining the learning rules corresponding to the objective, through which we can perform the exact optimization. Sometimes the learning rule may be complicated to derive, or sometimes the projection algorithm involves a great deal of computation or memory, which we may consider to be a drawback to overcome. The remainder of this book will be concerned with the development of some general algorithms with non-standard adaptation methods with this drawback in mind.

4

Non-standard Clustering Criteria

The K-means algorithm is one of the most frequently used investigatory algorithms in data analysis [108]. The algorithm attempts to locate K prototypes or means throughout a data set in such a way that the K prototypes in some way best represent the data.

In this book, we investigate alternative performance functions and show the effect the different functions have on the effectiveness of the resulting algorithms. We are specifically interested in developing algorithms which are effective in a worst case scenario: when the prototypes are initialized at the same position which is very far from the data points. This may initially sound an unlikely scenario but in a typical high dimensional space, most of the probability mass lies in the outer shell of the data [227]: so initialization to a mean of a set of data points may well lie closer to the centre than the individual data points and initialization to a single data point may cause distances to be measured across the empty centre of the space. These are well-known aspects of the "curse of dimensionality". If an algorithm can cope with these unfavourable scenarios, it should be able to cope with a more benevolent initialization.

We wish to overcome the problem of dependency on initial conditions by creating algorithms for prototype placement based on different performance functions.

4.1 A Family of New Algorithms

In this book, we introduce a new family of algorithms that solve the problem of sensitivity to initial conditions in the K-means and the soft K-means algorithms. The central idea in the new algorithms is that the change in position of each prototype is a function not just of its position with respect to the data but also takes into account the relative positions of all the other prototypes with respect to the data. Thus, if we have two clusters of data and two prototypes, one of them being closer to the first cluster, the other prototype

W.A. Barbakh et al.: Non-Stand. Param. Adapt. for Explor. Data Analys., SCI 249, pp. 49–72.
springerlink.com

will, before responding, recognize that there is one prototype closer to the first cluster and hence will prefer to move toward the second cluster.

4.1.1 Weighted K-Means Algorithm (WK)

We have previously investigated this effect in detail in [13, 14]. The performance function for the K-means may be written as

$$J_{Km} = \sum_{i=1}^{N} \min_{j=1}^{K} \| \mathbf{x}_i - \mathbf{m}_j \|^2 \tag{4.1}$$

which we wish to minimise by moving the prototypes to appropriate positions. Alternatively, we may write

$$J_{Km} = \sum_{i=1}^{N} \sum_{j=1}^{K} r_{ji} \| \mathbf{x}_i - \mathbf{m}_j \|^2 \tag{4.2}$$

where $r_{ji} = 1$ for $j = \arg\min_k \| \mathbf{x}_i - \mathbf{m}_k \|$ and 0 otherwise. Note that (4.1) detects only the prototypes closest to data points and then distributes them to give the minimum performance which determines the clustering. Any prototype which is still far from data is not utilised and does not enter any calculation that gives minimum performance. This may result in dead prototypes, which are never appropriate for any cluster. Thus initializing prototypes appropriately can have a big effect in the K-means algorithm.

We might consider the following performance function:

$$J_A = \sum_{i=1}^{N} \sum_{j=1}^{K} \| \mathbf{x}_i - \mathbf{m}_j \|^2 \tag{4.3}$$

which provides a relationship between all the data points and prototypes, but it doesn't provide useful clustering at minimum performance since

$$\frac{\partial J_A}{\partial \mathbf{m}_k} = 0 \Longrightarrow \mathbf{m}_k = \frac{1}{N} \sum_{i=1}^{N} \mathbf{x}_i, \forall k \tag{4.4}$$

Minimizing this performance function groups all the prototypes to the centre of data set regardless of the initial position of the prototypes which is useless for identification of clusters.

We wish to form a performance function with the following properties:

- Minimum performance gives an intuitively 'good' clustering.
- It creates a relationship between all data points and all prototypes.

(4.3) provides an attempt to reduce the sensitivity to prototypes' initialization by making a relationship between all data points and all prototypes while (4.1) provides an attempt to cluster data points at the minimum of the performance function. (4.1) is too local while (4.3) is too global. Therefore it may seem that what we want is to combine features of (4.1) and (4.3) to make a performance function such as:

$$J_{WK} = \sum_{i=1}^{N} \left[\sum_{j=1}^{K} \| \mathbf{x}_i - \mathbf{m}_j \| \right] \min_{k=1}^{K} \| \mathbf{x}_i - \mathbf{m}_k \|^2 \tag{4.5}$$

The rationale behind this performance function is that we wish to utilise the minimum distance in our learning algorithm but retain the global interaction which is necessary to ensure all prototypes play a part in the clustering. We derive the clustering algorithm associated with this performance function by calculating the partial derivatives of (4.5) with respect to the prototypes. We call the resulting algorithm Weighted K-means (though recognising that other weighted versions of K-means have been developed in the literature). Let

$$k* = \arg\min_{k=1}^{K}(\| \mathbf{x}_i - \mathbf{m}_k \|) \tag{4.6}$$

Then

$$\frac{\partial J_{WK,i}}{\partial \mathbf{m}_{k*}} = -(\mathbf{x}_i - \mathbf{m}_{k*})\{\| \mathbf{x}_i - \mathbf{m}_{k*} \| + 2\sum_{j=1}^{K} \| \mathbf{x}_i - \mathbf{m}_j \|\} = -(\mathbf{x}_i - \mathbf{m}_{k*})a_{ik*} \tag{4.7}$$

when $\mathbf{m}_{k*}$ is the closest prototype to $\mathbf{x}_i$ and

$$\frac{\partial J_{WK,i}}{\partial \mathbf{m}_j} = -(\mathbf{x}_i - \mathbf{m}_j)\frac{\| \mathbf{x}_i - \mathbf{m}_{k*} \|^2}{\| \mathbf{x}_i - \mathbf{m}_j \|} = -(\mathbf{x}_i - \mathbf{m}_j)b_{ij} \tag{4.8}$$

otherwise. Note that i determines $k*$, the closest prototype to $\mathbf{x}_i$ and hence the value of b_{ij}. We then solve this by summing over the whole data set and finding the fixed point solution of

$$\frac{\partial J_{WK}}{\partial \mathbf{m}_r} = \sum_{i=1}^{N} \frac{\partial J_{WK,i}}{\partial \mathbf{m}_r} = 0 \tag{4.9}$$

which gives a solution of

$$\mathbf{m}_r(t+1) = \frac{\sum_{i \in V_r} \mathbf{x}_i a_{ir} + \sum_{i \in V_j, j \neq r} \mathbf{x}_i b_{ir}}{\sum_{i \in V_r} a_{ir} + \sum_{i \in V_j, j \neq r} b_{ir}} \tag{4.10}$$

where V_r contains the indices of data points that are closest to $\mathbf{m}_r$, V_j contains the indices of all the other points and

$$a_{ir} = \| \mathbf{x}_i - \mathbf{m}_r(t) \| + 2\sum_{j=1}^{K} \| \mathbf{x}_i - \mathbf{m}_j \| \tag{4.11}$$

$$b_{ir} = \frac{\| \mathbf{x}_i - \mathbf{m}_{k*} \|^2}{\| \mathbf{x}_i - \mathbf{m}_r(t) \|} \tag{4.12}$$

where again $k* = \arg\min_j \parallel \mathbf{x}_i - \mathbf{m}_j \parallel$. We have given extensive simulations in [14] showing that this algorithm will cluster the data with the prototypes which are closest to the data points being positioned in such a way that the clusters can be identified. However there can be some potential prototypes which are not sufficiently responsive to the data and so never move to identify a cluster. In fact, these points move to a (weighted) centre of the data set. Practically, we have derived a better version of WK clustering algorithm by changing the exponential in the denominator of (4.12) from 1 to 2.

4.1.2 Inverse Weighted K-Means Algorithm (IWK)

Consider the performance function

$$J_{IWK} = \sum_{i=1}^{N} \left[\sum_{j=1}^{K} \frac{1}{\parallel \mathbf{x}_i - \mathbf{m}_j \parallel^p} \right] \min_{k=1}^{K} \parallel \mathbf{x}_i - \mathbf{m}_k \parallel^n \tag{4.13}$$

The rationale for this performance function is that we do not wish the situation to arise in which one prototype is optimally responsive to a data point but all the other prototypes are ignoring this data point: we wish all prototypes to take into account all data points. Let $\mathbf{m}_{k*}$ be the closest prototype to $\mathbf{x}_i$. Then

$$\begin{aligned} J_{IWK}(\mathbf{x}_i) &= \left[\sum_{j=1}^{K} \frac{1}{\parallel \mathbf{x}_i - \mathbf{m}_j \parallel^p} \right] \parallel \mathbf{x}_i - \mathbf{m}_{k*} \parallel^n \\ &= \parallel \mathbf{x}_i - \mathbf{m}_{k*} \parallel^{n-p} + \sum_{j \neq k*} \frac{\parallel \mathbf{x}_i - \mathbf{m}_{k*} \parallel^n}{\parallel \mathbf{x}_i - \mathbf{m}_j \parallel^p} \end{aligned} \tag{4.14}$$

Therefore the intuition behind this performance function is that it is minimised when the distance between the data and the closest prototype is minimal and the other prototypes are scattered widely throughout the data to give the largest possible value in the denominator of the second part. Also writing the performance function in this manner highlights the constraints which will be necessary on the values of p and n (see below). To create a learning rule, we calculate

$$\begin{aligned} \frac{\partial J_{IWK}(\mathbf{x}_i)}{\partial \mathbf{m}_{k*}} &= -(n-p)(\mathbf{x}_i - \mathbf{m}_{k*}) \parallel \mathbf{x}_i - \mathbf{m}_{k*} \parallel^{n-p-2} \\ &\quad -n(\mathbf{x}_i - \mathbf{m}_{k*}) \parallel \mathbf{x}_i - \mathbf{m}_{k*} \parallel^{n-2} \sum_{j \neq k*} \frac{1}{\parallel \mathbf{x}_i - \mathbf{m}_j \parallel^p} \\ &= (\mathbf{x}_i - \mathbf{m}_{k*}) a_{ik*} \end{aligned} \tag{4.15}$$

$$\frac{\partial J_{IWK}(\mathbf{x}_i)}{\partial \mathbf{m}_j} = p(\mathbf{x}_i - \mathbf{m}_j) \frac{\parallel \mathbf{x}_i - \mathbf{m}_{k*} \parallel^n}{\parallel \mathbf{x}_i - \mathbf{m}_j \parallel^{p+2}} = (\mathbf{x}_i - \mathbf{m}_j) b_{ij} \tag{4.16}$$

At convergence, $E(\frac{\partial J_{IWK}}{\partial \mathbf{m}_r}) = 0$ where the expectation is taken over the data set. With some abuse of our notation, we denote by V_j the set of points, $\mathbf{x}$ for which $\mathbf{m}_j$ is the closest, we have

$$\frac{\partial J_{IWK}}{\partial \mathbf{m}_r} = 0 \Longleftrightarrow \int_{\mathbf{x}\in V_r} \{(n-p)(\mathbf{x}-\mathbf{m}_{k*}) \parallel \mathbf{x}-\mathbf{m}_{k*} \parallel^{n-p-2}$$
$$+n(\mathbf{x}-\mathbf{m}_{k*}) \parallel \mathbf{x}-\mathbf{m}_{k*} \parallel^{n-2} \sum_{j\neq k} \frac{1}{\parallel \mathbf{x}-\mathbf{m}_j \parallel^p} P(\mathbf{x})\}d\mathbf{x}$$
$$+ \sum_{j\neq r} \int_{\mathbf{x}\in V_j} p(\mathbf{x}-\mathbf{m}_j) \frac{\parallel \mathbf{x}-\mathbf{m}_{k*} \parallel^n}{\parallel \mathbf{x}-\mathbf{m}_j \parallel^{p+2}} P(\mathbf{x})d\mathbf{x} = 0 \tag{4.17}$$

where $P(\mathbf{x})$ is the probability associated with the data set. This is, in general, a very difficult set of equations to solve. However it is readily seen that, for example, in the special case that there are the same number of prototypes as there are data points, that one solution is to locate each prototype at each data point (at which time $\frac{\partial J_{IWK}}{\partial \mathbf{m}_r} = 0$). Again solving this over all the data set results in

$$\mathbf{m}_r(t+1) = \frac{\sum_{i\in V_r} \mathbf{x}_i a_{ir} + \sum_{i\in V_j, j\neq r} \mathbf{x}_i b_{ir}}{\sum_{i\in V_r} a_{ir} + \sum_{i\in V_j, j\neq r} b_{ir}} \tag{4.18}$$

where V_r contains the indices of data points that are closest to $\mathbf{m}_r$, V_j contains the indices of all the other points and

$$a_{ir} = -(n-p) \parallel \mathbf{x}_i - \mathbf{m}_r(t) \parallel^{n-p-2}$$
$$-n \parallel \mathbf{x}_i - \mathbf{m}_r(t) \parallel^{n-2} \sum_{j\neq k*} \frac{1}{\parallel \mathbf{x}_i - \mathbf{m}_j \parallel^p} \tag{4.19}$$

$$b_{ir} = p \frac{\parallel \mathbf{x}_i - \mathbf{m}_{k*} \parallel^n}{\parallel \mathbf{x}_i - \mathbf{m}_r(t) \parallel^{p+2}} \tag{4.20}$$

From (4.17), we see that $n \geq p$ if the direction of the first term is to be correct and $n \leq p+2$ to ensure stability in all parts of that equation. In practice, we have found that a viable algorithm may be found by using (4.20) for all prototypes (and thus never using (4.19) for the closest prototype). We will call this the Inverse Weighted K-means algorithm (IWK) [181].

Discussion

Consider the performance function

$$J_{IWK} = \sum_{i=1}^{N} \left[\sum_{j=1}^{K} \frac{1}{\parallel \mathbf{x}_i - \mathbf{m}_j \parallel^p} \right] \min_{k=1}^{K} \parallel \mathbf{x}_i - \mathbf{m}_k \parallel^n \tag{4.21}$$

A priori, it may seem possible that from (4.21), a minimum of the performance function will be attained when at least some of the prototypes tend

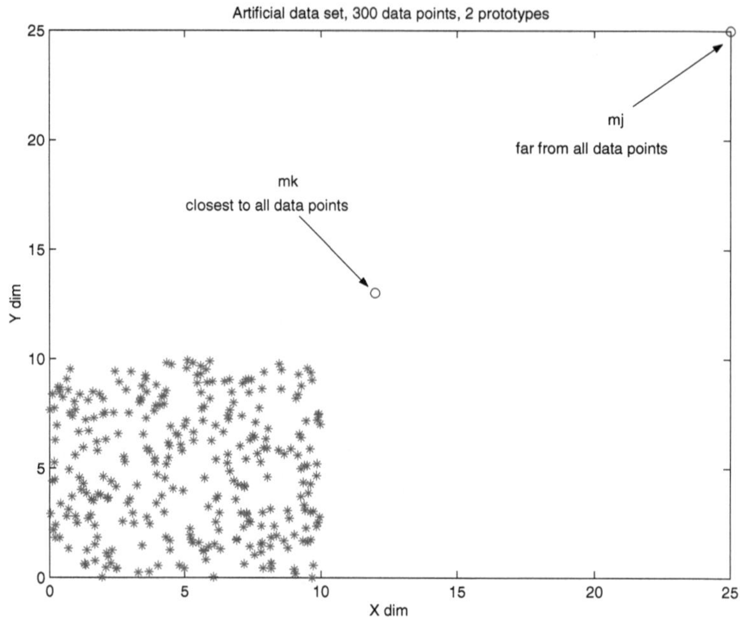

Fig. 4.1. Data points are shown as red '*'s. Prototypes are shown as blue 'o's.

to infinity. In fact, during the execution of the algorithm, sometimes one can see some prototypes initially move away from the data, but eventually one will see them returning to the centre of the data set.

We illustrate the general case with a toy example: let $\mathbf{m}_{k*}$ be the closest prototype to $\mathbf{x}_i$. Then

$$\begin{aligned} J_{IWK}(\mathbf{x}_i) &= \left[\sum_{j=1}^{K} \frac{1}{\| \mathbf{x}_i - \mathbf{m}_j \|^p} \right] \| \mathbf{x}_i - \mathbf{m}_{k*} \|^n \\ &= \| \mathbf{x}_i - \mathbf{m}_{k*} \|^{n-p} + \sum_{j \neq k*} \frac{\| \mathbf{x}_i - \mathbf{m}_{k*} \|^n}{\| \mathbf{x}_i - \mathbf{m}_j \|^p} \end{aligned} \tag{4.22}$$

Let us have an artificial data set and two prototypes as shown in Figure 4.1. Let us consider the case of $n = p + 2$ $(p = 1)$, and any $\mathbf{x}_i$, for example $\mathbf{x}_i = (10, 10)$. If $\mathbf{m}_j$ goes to infinity and $\mathbf{m}_{k*}$ moves to the mean of the data, then:

$\mathbf{m}_k = (5, 5)$, $\mathbf{m}_j = (\infty, \infty)$

$$J_{IWK}(\mathbf{x}_i) = \left[\sum_{j=1}^{K} \frac{1}{\| \mathbf{x}_i - \mathbf{m}_j \|^p} \right] \| \mathbf{x}_i - \mathbf{m}_{k*} \|^n$$

$$= \| \mathbf{x}_i - \mathbf{m}_{k*} \|^{n-p} + \sum_{j \neq k*} \frac{\| \mathbf{x}_i - \mathbf{m}_{k*} \|^n}{\| \mathbf{x}_i - \mathbf{m}_j \|^p}$$

$$= 7.0711^2 + \frac{7.0711^3}{\infty}$$

$$= 50$$

The performance function value for the whole data set shown in the figure has been calculated as $J_{IWK} = 17813$.

Figure 4.2 shows the result obtained after executing IWK.

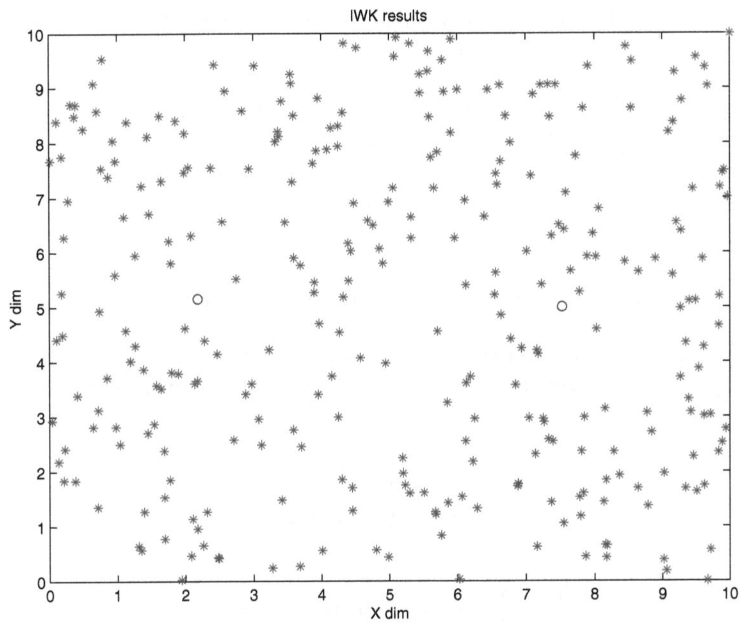

Fig. 4.2. IWK results

Consider still the point (10,10). Then with the IWK, $\mathbf{m}_1 = (2.1825, 5.1579)$, $\mathbf{m}_2 = (7.5290, 5.0057)$, the values from Figure 4.2

$$J_{IWK}(\mathbf{x}_i) = \left[\sum_{j=1}^{K} \frac{1}{\| \mathbf{x}_i - \mathbf{m}_j \|^p} \right] \| \mathbf{x}_i - \mathbf{m}_{k*} \|^n$$

$$= \| \mathbf{x}_i - \mathbf{m}_{k*} \|^{n-p} + \sum_{j \neq k*} \frac{\| \mathbf{x}_i - \mathbf{m}_{k*} \|^n}{\| \mathbf{x}_i - \mathbf{m}_j \|^p}$$

$$= 5.5722^2 + \frac{5.5722^3}{9.1956}$$

$$= 49.8642$$

The performance function value for the whole data set in the figure has been calculated as $J_{IWK} = 5150.1$, i.e. considerably lower than the first case.

So from this example, we see that it is less important to have the prototypes far from data points than it is to have the prototypes centered on the data. Also, if we use more than two prototypes, we find that the performance function value decreases when these prototypes move to fit the data, while the infinity solution stays high(see the results in Table 4.1).

Table 4.1. First column: the number of prototypes. Second column: the performance function value when one prototype moves to the mean of the data and the rest go to ∞. Third column: the performance function value when all the prototypes are located using IWK algorithm.

Number of prototypes	J_{IWK}	J_{IWK}
2	$J_{IWK} = 17813$	$J_{IWK} = 5150.1$
3	$J_{IWK} = 17813$	$J_{IWK} = 3823.6$
4	$J_{IWK} = 17813$	$J_{IWK} = 2720.3$

4.1.3 The Inverse Weighted Clustering Algorithm

The general drive towards the data (and away from any solution which has prototypes tending towards ∞) is given by (4.20) Consider the following performance function:

$$J_I = \sum_{i=1}^{N} \sum_{k=1}^{K} \frac{1}{\| \mathbf{x}_i - \mathbf{m}_k \|^P} \tag{4.23}$$

$$\frac{\partial J_I}{\partial \mathbf{m}_k} = \sum_{i=1}^{N} P(\mathbf{x}_i - \mathbf{m}_k) \frac{1}{\| \mathbf{x}_i - \mathbf{m}_k \|^{P+2}} \tag{4.24}$$

$$\frac{\partial J_I}{\partial \mathbf{m}_k} = 0 \Longrightarrow \mathbf{m}_k = \frac{\sum_{i=1}^{N} b_{ik} \mathbf{x}_i}{\sum_{i=1}^{N} b_{ik}}$$

where

$$b_{ik} = \frac{1}{\| \mathbf{x}_i - \mathbf{m}_k \|^{P+2}} \tag{4.25}$$

The partial derivative of J_I with respect to $\mathbf{m}_k$ can be used to maximize the performance function J_I. So the implementation of (4.25) will always move $\mathbf{m}_k$ to the closest data point to maximize J_I to ∞.

However, the implementation of (4.25) will not identify any clusters as the prototypes always move to the closest data point. But the advantage of this performance function is that it does not leave any prototype far from data: all the prototypes converge toward the data.

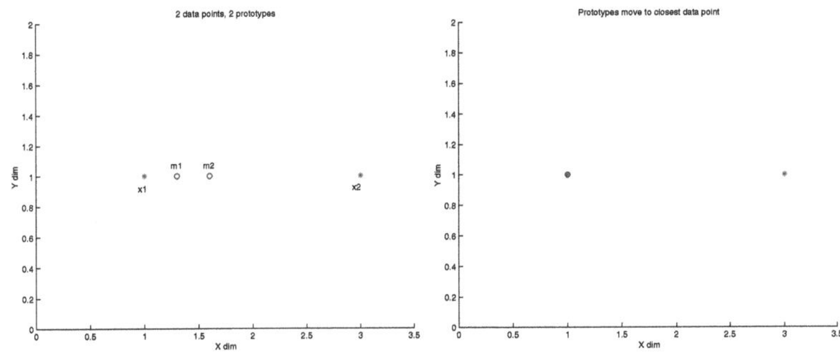

Fig. 4.3. Left: two data points and two prototypes. Right: the result after applying (4.25).

We can enhance this algorithm's ability to identify the clusters without losing its property of pushing the prototypes inside data clusters by changing b_{ik} in (4.25) to the following:

$$b_{ik} = \frac{\| \mathbf{x}_i - \mathbf{m}_{k*} \|^{P+2}}{\| \mathbf{x}_i - \mathbf{m}_k \|^{P+2}} \tag{4.26}$$

where $\mathbf{m}_{k*}$ is the closest prototype to $\mathbf{x}_i$.

With this change, we have an interesting behavior: (4.26) works to maximize J_I by moving the prototypes to the free data points (or clusters) instead of the closest data point (or local cluster).

Note that (4.26) is derived from the following performance function:

$$J_{IWC} = \sum_{i=1}^{N} \left[\sum_{j \neq k*}^{K} \frac{1}{\| \mathbf{x}_i - \mathbf{m}_j \|^P} \right] * \left(\| \mathbf{x}_i - \mathbf{m}_{k*} \|^{P+2} \right) \tag{4.27}$$

where

$$k* = \arg \min_{k=1}^{K} (\| \mathbf{x}_i - \mathbf{m}_k \|)$$

We will call this the Inverse Weighted Clustering algorithm (IWC) [16], which is a special case of the inverse weighted K-means (IWK) algorithm.

Note that (4.25) and (4.26) never leave any prototype far from the data even if they are initialized outwith the data. The prototypes are always pushed to join the closest data points using (4.25) or to join the free data points using (4.26). But (4.25) doesn't identify clusters while (4.26) does: (4.26) keeps the property of (4.25) of pushing the prototypes to join the data, and provides the ability of identifying clusters.

Consider we have two data points and two prototypes, so we have the following possibilities:

1. Two prototypes are closest to one data point, as shown in Figure 4.3, left.
2. One prototype is closest only to one data point, as shown in Figure 4.4.
3. One prototype is closest to both data points, as shown in Figure 4.5.

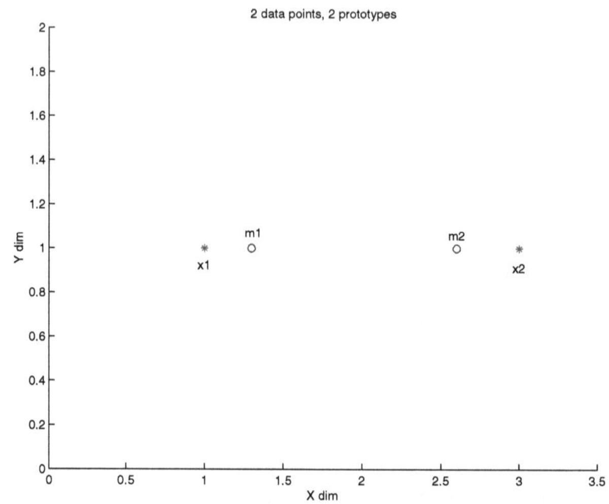

Fig. 4.4. One prototype is closest only to one data point

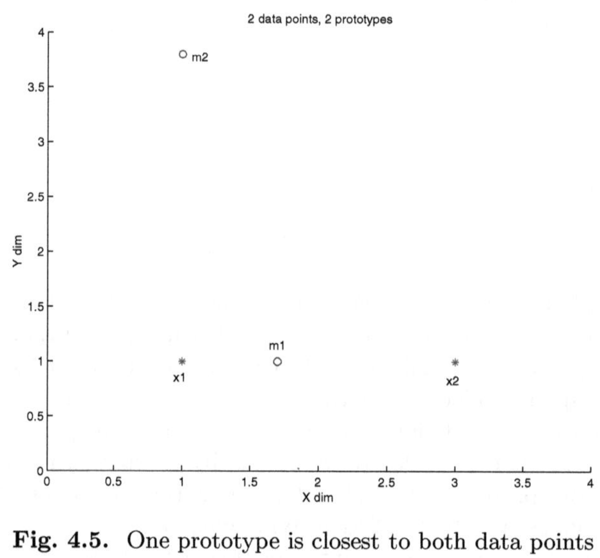

Fig. 4.5. One prototype is closest to both data points

Analysis for first possibility
With (4.25),

$$\mathbf{m}_1 = \frac{\frac{1}{d_{11}^{(P+2)}}\mathbf{x}_1 + \frac{1}{d_{21}^{(P+2)}}\mathbf{x}_2}{\frac{1}{d_{11}^{(P+2)}} + \frac{1}{d_{21}^{(P+2)}}} \tag{4.28}$$

where

$$d_{ik} = \parallel \mathbf{x}_i - \mathbf{m}_k \parallel$$

if $d_{11} < d_{21}$ ($\mathbf{m}_1$ is closer to $\mathbf{x}_1$)
$\mathbf{m}_1$ will move toward $\mathbf{x}_1$
else if $d_{11} > d_{21}$ ($\mathbf{m}_1$ is closer to $\mathbf{x}_2$)
$\mathbf{m}_1$ will move toward $\mathbf{x}_2$
else ($\mathbf{m}_1$ is located at the mean of data)
$\mathbf{m}_1$ will remain at the mean of data

Similarly for the prototype $\mathbf{m}_2$: $\mathbf{m}_2$ will move independently toward the closest data point without taking into account how the other prototypes respond. There is no way to identify clusters using (4.25).

With (4.26), b_{ik} is always in the range [0 1].

$$\mathbf{m}_1 = \frac{\frac{d_{11}^{(P+2)}}{d_{11}^{(P+2)}}\mathbf{x}_1 + \frac{d_{22}^{(P+2)}}{d_{21}^{(P+2)}}\mathbf{x}_2}{\frac{d_{11}^{(P+2)}}{d_{11}^{(P+2)}} + \frac{d_{22}^{(P+2)}}{d_{21}^{(P+2)}}}$$

Normally $\frac{d_{22}^{(P+2)}}{d_{21}^{(P+2)}} < 1$, and $\mathbf{m}_1$ will move toward $\mathbf{x}_1$.
(If this value =1, then $\mathbf{m}_1$ will move to the mean.)

$$\mathbf{m}_2 = \frac{\frac{d_{11}^{(P+2)}}{d_{12}^{(P+2)}}\mathbf{x}_1 + \frac{d_{22}^{(P+2)}}{d_{22}^{(P+2)}}\mathbf{x}_2}{\frac{d_{11}^{(P+2)}}{d_{12}^{(P+2)}} + \frac{d_{22}^{(P+2)}}{d_{22}^{(P+2)}}} \tag{4.29}$$

Normally $\frac{d_{11}^{(P+2)}}{d_{12}^{(P+2)}} < 1$, and $\mathbf{m}_2$ will move toward $\mathbf{x}_2$, although $\mathbf{m}_2$ is closer to $\mathbf{x}_1$.

Notice, if we have two prototypes, one initialized at the mean and the second initialized anywhere between the two data points, we will find that the second prototype is closer to one data point and hence after the next iteration each data point will move towards a data point. So there is no problem if any prototype moves toward the mean.

Analysis for second possibility
(4.25) and (4.26) give the same effect. Each prototype will move toward the closest data point.

Analysis for third possibility

With (4.25), each prototype moves to the closest data point, so for Figure 4.5, $\mathbf{m}_1$ and $\mathbf{m}_2$ will move to the same data point (1,1).

With (4.26), after the first iteration, $\mathbf{m}_1$ will move to the mean of data as it is the closest for both data points, and $\mathbf{m}_2$ will move to a location between the two data points and then we get the first or second possibility for the next iteration.

$$\mathbf{m}_1 = \frac{\frac{d_{11}^{(P+2)}}{d_{11}^{(P+2)}}\mathbf{x}_1 + \frac{d_{21}^{(P+2)}}{d_{21}^{(P+2)}}\mathbf{x}_2}{\frac{d_{11}^{(P+2)}}{d_{11}^{(P+2)}} + \frac{d_{21}^{(P+2)}}{d_{21}^{(P+2)}}}$$

$$\mathbf{m}_2 = \frac{\frac{d_{11}^{(P+2)}}{d_{12}^{(P+2)}}\mathbf{x}_1 + \frac{d_{21}^{(P+2)}}{d_{22}^{(P+2)}}\mathbf{x}_2}{\frac{d_{11}^{(P+2)}}{d_{12}^{(P+2)}} + \frac{d_{21}^{(P+2)}}{d_{22}^{(P+2)}}}$$

Simulations

In Figure 4.6, the prototypes have all been initialized within a single cluster. As shown in the figure, while the K-means failed to identify clusters, bottom

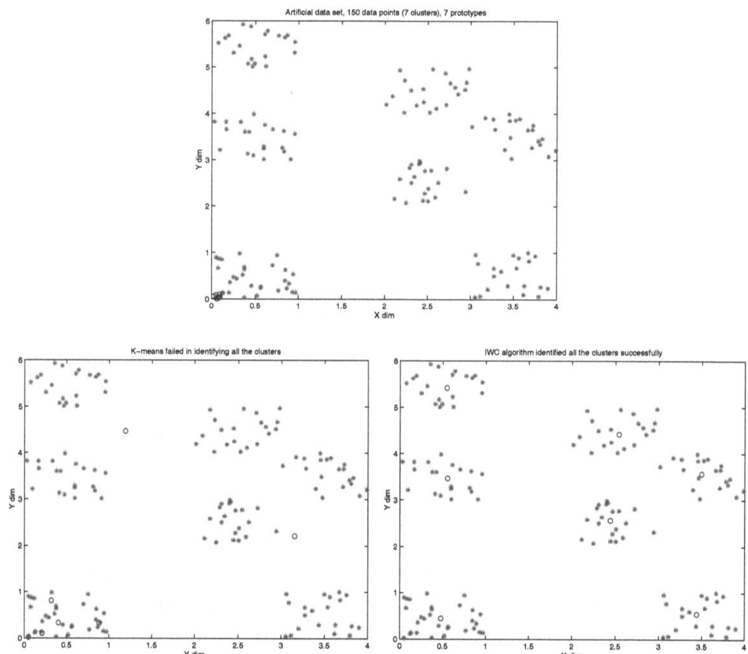

Fig. 4.6. Top: Artificial data set: the data set is shown as 7 clusters of red '*'s; the prototypes are initialized to lie within one cluster and shown as blue 'o's. Bottom left: the K-means result. Bottom right: the IWC algorithm result.

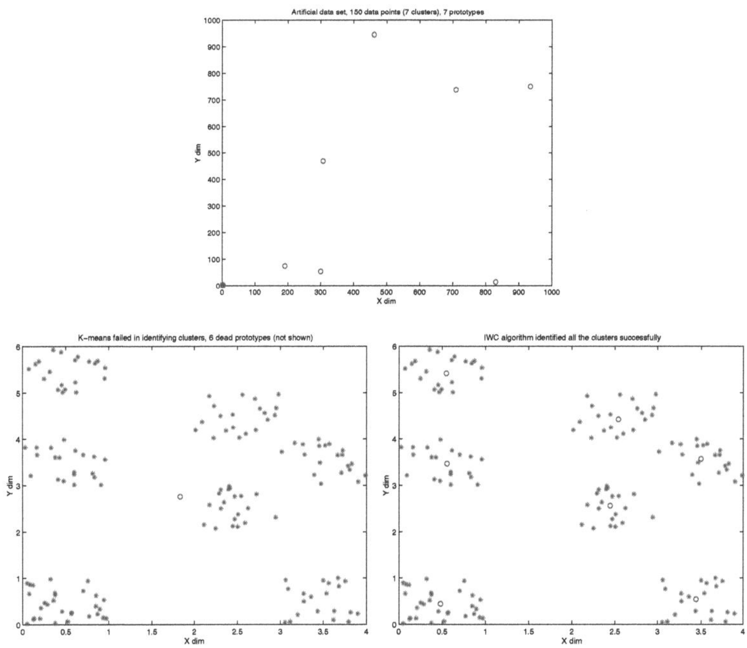

Fig. 4.7. Top: Artificial data set: the data set is shown as 7 clusters of red '*'s; the prototypes are initialized very far from data and shown as blue 'o's. Bottom left: the K-means result. Bottom right: the IWC algorithm result.

left, the IWC based on (4.26) identified all of them successfully, bottom right. Note that in this simulation, some centres for the K-means did not have enough impetus to leave the data for which they already had responsibility whereas with the IWC there is a strong drive to get centres to take high responsibility for all data.

Figure 4.7 shows the result of applying the IWC algorithm to the same artificial data set but with poor initialization of the prototypes. As shown in the figure, the inverse weighted clustering algorithm (IWC) succeeds in identifying the clusters under this poor initialization, bottom right, while the K-means failed, bottom left. Note that with this simulation, some prototypes in the K-means remain dead prototypes while this does not happen in the IWC algorithm.

The two effects of the last two groups of the K-means simulations - not moving from data already captured and leaving some prototypes as dead prototypes - appear in all subsequent simultions in this book. All new methods in this book overcome both of these difficulties though we will often show just a simulation exhibiting one or other behaviour rather than both.

In general initializing prototypes far from data is an unlikely situation to happen, but it may be that all the prototypes are in fact initialized very far

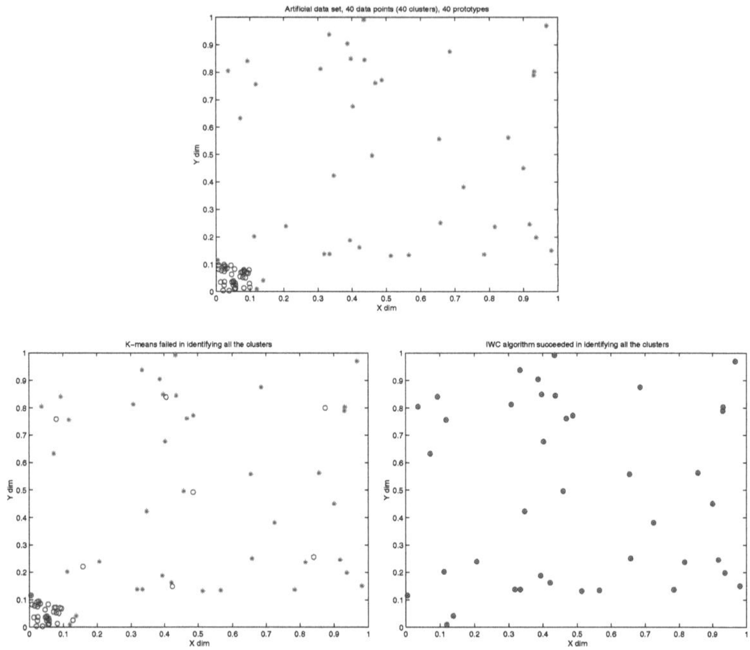

Fig. 4.8. Top: Artificial data set: the data set is shown as 40 clusters of red '*'s; 40 prototypes are initialized close together and shown as blue 'o's. Bottom left: the K-means result. Bottom right: the IWC algorithm result.

from a particular cluster. In that situation, the algorithms in this book will identify the cluster while the K-means may not.

In Figure 4.8, we have 40 data points, each of which represents one cluster. 40 prototypes are initialized very close together. The IWC algorithm, bottom right, gives a better result than the K-means algorithm, bottom left.

From extensive simulations, we can confirm that (4.26) always pushes the prototypes toward the data.

In the IWK algorithm, the second equation (4.20) is exactly similar to that one we analyzed above. So, the IWK always searches for the minimum performance function by pushing the prototypes in such a way inside data to identify clusters. Even if the prototypes are initialized far from data, the IWK will move them inside data clusters and will not leave any of them outside. The IWK will search only for the minimum performance that distributes all the prototypes inside the data thus identifying clusters.

In Figure 4.9, the prototypes have all been initialized within a single cluster. As shown in the Figure, while the K-means algorithm failed to identify the clusters, the new algorithm IWK identified them successfully.

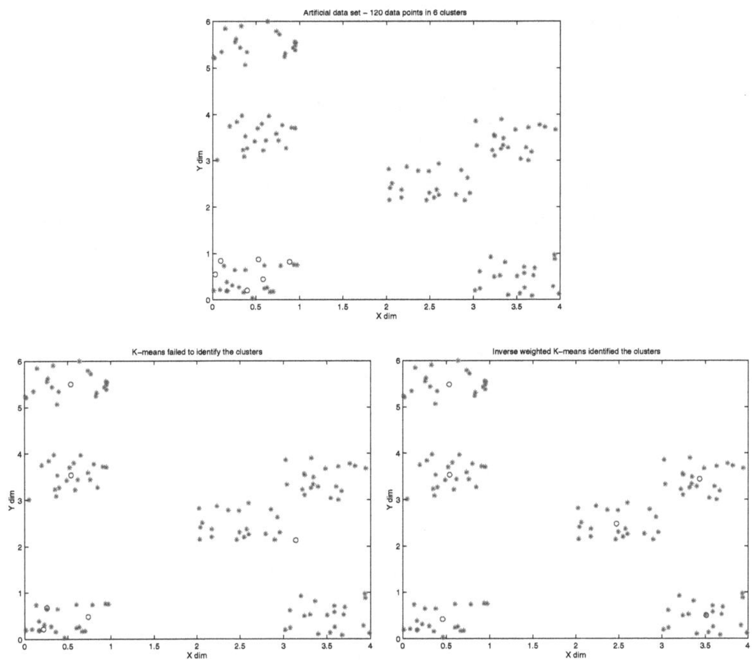

Fig. 4.9. Top: Artificial data set, data set is shown as 6 clusters of red '*'s, prototypes are initialized to lie within one cluster and shown as blue 'o's. Bottom left: the K-means result. Bottom right: the IWK result.

Relation to Other Models

Recently [69] discussed a 'rotational invariant L_1-norm principal component analysis'. They point out that PCA typically minimises

$$J_{PCA} = \sum_{i=1}^{N} \parallel \mathbf{x}_i - (\mathbf{x}_i^T \mathbf{w})\mathbf{w} \parallel^2 \tag{4.30}$$

while robust L_1-norm PCA can be defined as the minimisation of

$$J_{L_1} = \sum_{i=1}^{N} \parallel \mathbf{x}_i - (\mathbf{x}_i^T \mathbf{w})\mathbf{w} \parallel_1 \tag{4.31}$$

where the last subscript identifies the L_1 norm. They note that there is an inherent assumption that the summation over the samples, subscripted i are treated the same as the summation over each data sample. This is most clearly seen by writing (4.30) as

$$J_{PCA} = \sum_{i=1}^{N} \sum_{d=1}^{D} (x_{id} - (\mathbf{x}_i^T \mathbf{w})w_d)^2 \tag{4.32}$$

where D is the dimensionality of the data. They denote

$$J_{R_1} = \sum_{i=1}^{N} [\sum_{d=1}^{D} (x_{id} - (\mathbf{x}_i^T \mathbf{w}) w_d)^2]^{\frac{1}{2}} \tag{4.33}$$

the performance function for their rotational invariant L_1-norm principal component analysis though it is clearly a mixture of L_1 and L_2 norms.

Now we note that we may use the same function as a norm for the K-means. Thus

$$\begin{aligned} J_{R_1} &= \sum_{i=1}^{N} \min_{j=1}^{K} [\sum_{d=1}^{D} (x_{id} - m_{jd})^2]^{\frac{1}{2}} \\ &= \sum_{i=1}^{N} \min_{j=1}^{K} \| \mathbf{x}_i - \mathbf{m}_j \| \end{aligned}$$

i.e. a sum of L_2 norms rather than a sum of their squares.

We may show that this leads to the adaptation

$$\begin{aligned} \mathbf{m}_j &= \frac{\sum_{i \in V_j} b_{ij} \mathbf{x}_i}{\sum_{i \in V_j} b_{ij}} \\ \text{where } b_{ij} &= \frac{1}{\| \mathbf{x}_i - \mathbf{m}_j \|} \end{aligned} \tag{4.34}$$

This algorithm is very closely related to (4.25) in which $p = -1$.

4.1.4 Inverse Exponential K-Means Algorithm 1 (IEK1)

Of course there are other ways we can combine local and global knowledge in clustering algorithms. For example, we have provided a new algorithm called Inverse Exponential K-means algorithm [11, 17, 19] with the following performance function:

$$\begin{aligned} k* &= \arg \min_{k=1}^{K} (\| \mathbf{x}_i - \mathbf{m}_k \|) \\ J_{IEK} &= \sum_{i=1}^{N} \left[\sum_{j \neq k*}^{K} \frac{1}{\| \mathbf{x}_i - \mathbf{m}_j \|} \right] * \left(1 - \exp(- \| \mathbf{x}_i - \mathbf{m}_{k*} \|^3)\right) \end{aligned} \tag{4.35}$$

This performance function deals with the prototypes that are not detected by the minimum function and hence it solves the problem of sensitivity to initial conditions in the K-means algorithm. We can therefore combine this performance function with the standard K-means performance function (4.1) to create an algorithm which will enable the derivation of two sets of update rules, as we will see in the optimization and implementation section. We use

(4.1) when $\mathbf{m}_j$ is the closest to $\mathbf{x}_i$ and use (4.35) for the other prototypes that are not the closest to $\mathbf{x}_i$.

The performance function (4.35) deals with all prototypes that are not recognized as closest to the data sample. This function at minimum values tries to distribute the prototypes to fit the data and find the clusters. We need now to generate a new algorithm by optimizing the performance functions to give the minimum value and hence identifying the clusters.

Optimization and Implementation

To derive the new clustering algorithm, we need to find the partial derivative of (4.1) with respect to $\mathbf{m}_k$ and the partial derivative of (4.35) with respect to $\mathbf{m}_j$ where $\mathbf{m}_k$ represents the closest prototype to $\mathbf{x}_i$, and $\mathbf{m}_j$ represents the other prototypes.

$$\frac{\partial J_{Km}}{\partial \mathbf{m}_k} = \sum_{i \in V_k} -2(\mathbf{x}_i - \mathbf{m}_k) \tag{4.36}$$

where V_k contains the indices of data points that are closest to $\mathbf{m}_k$. This leads to

$$\frac{\partial J_{Km}}{\partial \mathbf{m}_k} = 0$$

$$\mathbf{m}_k = \frac{\sum_{i \in V_k} \mathbf{x}_i}{N_r} = \frac{\sum_{i \in V_k} \mathbf{x}_i a_{ik}}{\sum_{i \in V_k} a_{ik}}$$

where N_r is the number of data points that are closest to $\mathbf{m}_k$, $a_{ik} = 1$.

Note that this constitutes only a part of the calculation of $\mathbf{m}_k$ from only the closest data points, however there is another calculation for $\mathbf{m}_k$ (using other data points). This is provided from the second performance function as $\mathbf{m}_k$ might not be the closest to some data points $\mathbf{x}_i$, $i \in V_j$ where V_j is the index of data points that are not closest to $\mathbf{m}_k$, see (4.40). Thus $\mathbf{m}_k$ is calculated based on all the data points, not only the closest points as happens in the K-means algorithm.

The second performance function provides new calculations for the prototypes that are not closest to data points, and distributes them appropriately to identify the clusters.

$$\begin{aligned}\frac{\partial J_{IEK}}{\partial \mathbf{m}_j} &= \sum_{i \in V_d} \frac{(\mathbf{x}_i - \mathbf{m}_j)(1 - \exp(- \parallel \mathbf{x}_i - \mathbf{m}_{k*} \parallel^3))}{\parallel \mathbf{x}_i - \mathbf{m}_j \parallel^3} \\ &= \sum_{i \in V_d} \frac{(\mathbf{x}_i - \mathbf{m}_j) c_i}{\parallel \mathbf{x}_i - \mathbf{m}_j \parallel^3}\end{aligned} \tag{4.37}$$

where V_d contains the indices of data points that are not closest to $\mathbf{m}_j$

$$c_i = 1 - \exp(- \parallel \mathbf{x}_i - \mathbf{m}_{k*} \parallel^3) \tag{4.38}$$

By assigning the partial derivative to zero and solving for $\mathbf{m}_j$ we have:

$$\frac{\partial J_{IEK}}{\partial \mathbf{m}_j} = 0$$

$$\mathbf{m}_j(t+1) = \frac{\sum_{i \in V_d} \frac{\mathbf{x}_i c_i}{\|\mathbf{x}_i - \mathbf{m}_j(t)\|^3}}{\sum_{i \in V_d} \frac{c_i}{\|\mathbf{x}_i - \mathbf{m}_j(t)\|^3}}$$

$$\text{or } \mathbf{m}_j(t+1) = \frac{\sum_{i \in V_d} \mathbf{x}_i b_{ij}}{\sum_{i \in V_d} b_{ij}} \tag{4.39}$$

where

$$b_{ij} = \frac{c_i}{\| \mathbf{x}_i - \mathbf{m}_j(t) \|^3} = \frac{1 - \exp(- \| \mathbf{x}_i - \mathbf{m}_{k*} \|^3)}{\| \mathbf{x}_i - \mathbf{m}_j(t) \|^3}$$

The new locations for all prototypes can be calculated by:

$$\mathbf{m}_r(t+1) = \frac{\sum_{i \in V_r} \mathbf{x}_i a_{ir} + \sum_{i \in V_j, j \neq r} \mathbf{x}_i b_{ir}}{\sum_{i \in V_r} a_{ir} + \sum_{i \in V_j, j \neq r} b_{ir}} \tag{4.40}$$

where V_r contains the indices of data points that are closest to $\mathbf{m}_r$, V_j contains the indices of all the other points and

$$a_{ir} = 1 \tag{4.41}$$

$$b_{ir} = \frac{1 - \exp(- \| \mathbf{x}_i - \mathbf{m}_{k*} \|^3)}{\| \mathbf{x}_i - \mathbf{m}_r(t) \|^3} \tag{4.42}$$

In execution we have found that the main computational cost is based on (4.42). Assume we have K prototypes and N data points; then as every data point is closest to one and only one prototype, for one iteration we have $N * K$ loops, and in each loop either (4.41) or (4.42) will be executed. (4.41) will be executed N times. (4.42) will be executed H times, where $H = N * (K - 1)$.

Thus it is possible now to find a viable algorithm by using (4.42) for all prototypes (and thus never using (4.41) for the closest prototype).

However, (4.41) provides a strong advantage by introducing two sets of updates rather than a single update for all prototypes. Therefore, this allows the algorithm to separate the joint prototypes and hence distribute them to give good clustering. This is a symmetry-breaking factor which enables local minima to be avoided and allows the algorithm to be superior under some difficult conditions.

We assume every data point has only one minimum distance to the prototypes, in other words it is closest to one prototype only. We treat the other prototypes as distant prototypes even if they have the same minimum value. This step is optional, but it is very important to allow the algorithm to work very well in the case that all prototypes are initialized in the same location.

This allows processing joint prototypes by two different equations and hence separates them.

4.1.5 Inverse Exponential K-Means Algorithm 2 (IEK2)

This algorithm has a similar rationale to the previous one. The new locations for the prototypes can be calculated using the following equation:

$$\mathbf{m}_r(t+1) = \frac{\sum_{i \in V_r} \mathbf{x}_i a_{ir} + \sum_{i \in V_j, j \neq r} \mathbf{x}_i b_{ir}}{\sum_{i \in V_r} a_{ir} + \sum_{i \in V_j, j \neq r} b_{ir}} \tag{4.43}$$

where V_r contains the indices of data points that are closest to $\mathbf{m}_r$, V_j contains the indices of all the other points and

$$a_{ir} = \sum_k \frac{1}{\exp(-d(\mathbf{x}_i, \mathbf{m}_k)^{\zeta})}$$

$$b_{ir} = \frac{(1 - \exp(-(\min_{k=1}^{K}(d(\mathbf{x}_i, \mathbf{m}_k)))^3))^3}{(1 + \epsilon - \exp(-d(\mathbf{x}_i, \mathbf{m}_r(t))^3))^3}$$

where typically $\zeta = 0.01$ and $\epsilon = 0.001$. This algorithm provides a good way to distribute the prototypes to fit the manifold a little bit better than IEK1.

4.1.6 Simulations

We illustrate these algorithms with a few simulations on artificial two dimensional data sets, since the results are easily verified visually. Consider now the situation in which the prototypes are initialized very far from the data. In Figure 4.10 we have 40 data points, each of which represents one cluster. All the prototypes are initialized in the same location and far from the clusters.

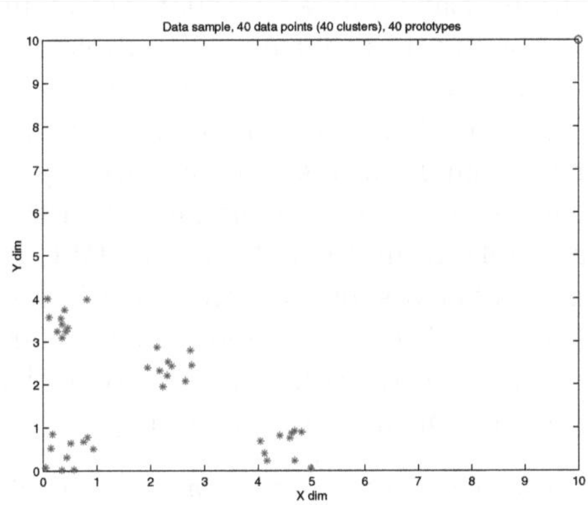

Fig. 4.10. A poor initialisation of the prototypes' locations: the prototypes are located at (10,10)

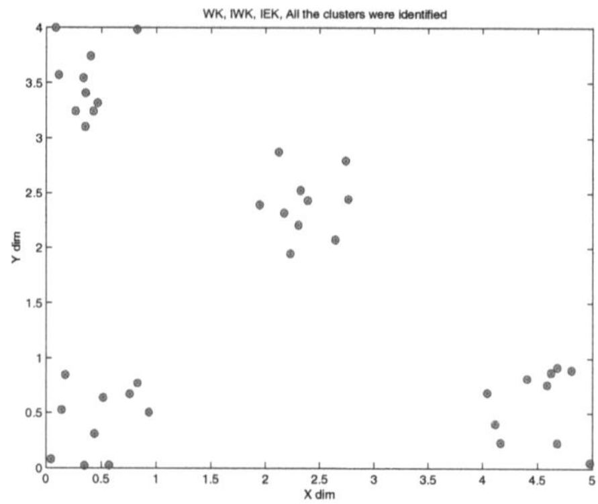

Fig. 4.11. Typical simulation: the resulting prototypes' positions after applying one of WK, IWK, IEK1 and IEK2

Figure 4.11 shows the result after applying any one of WK, IWK, IEK1 and IEK2: it is clear that under a very bad initialization the new algorithms still work well and succeed in identifying all the clusters. We haven't applied IWC to this artificial data set as it has only one set of update, and thus it can not separate the joint prototypes.

The iris[1] data set is a real data set with 150 random samples of flowers from the iris species setosa, versicolor, and virginica collected by Anderson in 1935 [6]. There are 50 observations from each species for sepal length, sepal width, petal length and petal width in cm. This data set was used by Fisher(1936) [79] in his initiation of the linear discriminant-function technique.

The glass[2] data set is another real data set which has 214 samples with 10 dimensions and 6 types.

We now apply K-means, WK, IWC, IWK, and IEK1 algorithms to the the iris and glass real data set. We run each algorithm with 10 different prototypes' initializations. Each time we calculate the quantization error (Qe) and the number of errors after clustering (Er). Table 4.2 shows the results on iris data set, and Table 4.3 shows the results on glass data set.

From Table 4.2 and Table 4.3, we see the new algorithms work better and more robustly in general than K-means and give better quantization error and lower number of errors. Also, we see all the new algorithms are comparable to each other. For example, from Table 4.2 we see IWC is the best in terms of the minimum number of errors, but it is the worst in terms of quantization error. However in Table 4.3, IWC gives a good result in terms of quantization error. Thus we can not say this algorithm is better than that; any one of them can be the best, depending on the data set we use.

[1] The iris data set is available at `http://mlearn.ics.uci.edu/databases/`.

[2] The glass data set is available at `http://mlearn.ics.uci.edu/databases/`.

Table 4.2. Results of applying K-means, IWC, WK, IWK and IEK1 to the iris data set. The rows show the results of applying the algorithms with different prototypes' initialization. Last row shows the average of quantization error and number of errors for each algorithm. For each algorithm, we have two columns, the first one shows the quantization error, and the second one shows the number of errors.

	K-means		IWC		WK		IWK		IEK1	
	Qe	Er	Qe	Er	Qe	Er	Qe	Er	Qe	Er
1	97.3	16	98.5	15	97.8	17	99.2	18	97.5	15
2	97.3	17	98.5	15	97.8	17	97.1	16	97.5	15
3	123.8	50	98.5	15	97.8	17	97.8	15	97.6	16
4	97.3	17	98.5	15	97.8	17	97.1	16	97.6	16
5	124.1	50	98.5	15	97.8	17	97.1	16	97.6	16
6	97.3	16	98.5	15	97.8	17	99.2	18	97.5	15
7	124.1	50	98.5	15	97.8	17	97.7	15	97.5	15
8	97.3	17	98.5	15	97.8	17	97.1	16	97.5	15
9	124.1	50	98.5	15	97.8	17	97.8	15	97.6	16
10	123.8	50	98.5	15	97.8	17	97.7	15	97.6	16
Avg.	110.6	33.3	98.5	15	97.8	17	97.78	16	97.55	15.5

Table 4.3. Results of applying K-means, IWC, WK, IWK and IEK1 to the glass data set. The rows show the results of applying the algorithms with different prototypes' initialization. Last row shows the average of quantization error and number of errors for each algorithm. For each algorithm, we have two columns, the first one shows the quantization error, and the second one shows the number of errors.

	K-means		IWC		WK		IWK		IEK1	
	Qe	Er	Qe	Er	Qe	Er	Qe	Er	Qe	Er
1	1977.3	23	1973.7	20	1972.1	19	1999.2	23	1979.6	25
2	1990.7	27	1973.5	19	1974.1	20	1989.1	22	1979.6	25
3	1979.6	25	1973.7	20	1973.2	21	1996.4	23	1977.1	21
4	1989.8	26	1973.7	20	1974.1	20	1985.0	22	1990.7	27
5	5761.8	75	1973.7	20	1971.8	20	1999.9	24	1974.6	20
6	1990.7	27	1973.7	20	1973.2	21	1990.4	23	1979.6	25
7	1979.6	25	1973.5	19	1974.1	20	1992.1	22	1979.6	25
8	1979.6	25	1973.5	19	1974.1	20	1994.6	23	1990.7	27
9	5761.8	75	1973.7	20	1974.1	20	1988.6	21	1977.1	21
10	1990.7	27	1973.7	20	1974.1	20	1999.2	23	1979.6	25
Avg.	2740.2	35.5	1973.6	19.7	1973.5	20.1	1993.5	22.6	1980.8	24.1

4.1.7 Summary

We have introduced new algorithms above as a family of algorithms. They were all developed with the same underlying rationale: they are designed to improve upon K-means' convergence to local optima depending on the initial conditions with which we start the algorithm. Also each of the algorithms incorporates both global and local knowledge so that the means when being re-positioned take account of the positions of all the other means and the data samples too.

In the weighted K-means algorithm (WK), it is possible on occasion to see some of the prototypes move to the centre of a data set and thus no clusters can be identified with these prototypes; they are essentially dead prototypes. The Inverse Weighted K-means (IWK) on the other hand, never shows such behaviour. In every single experiment we have performed with this algorithm, the prototypes have been sensibly located within whatever data set we were using.

The inverse exponential weighted K-means algorithms (IEK1 and IEK2) exhibit very similar behaviour to the inverse weighted K-means in that all prototypes are sensibly located. These algorithms may be thought to use a similar technique to the standard soft K-means [156] though the inverse exponential weighted K-means is much better at solving the problem of sensitivity to initial conditions [11].

All the algorithms, except the inverse weighted clustering algorithm (IWC) which is special case of IWK, have two sets of updates rather than a single update for all prototypes. This is a symmetry-breaking factor which enables local minima to be avoided, even if all prototypes are initialized in the same location as one point.

4.2 Spectral Clustering Algorithm

In this section we show how to use any one of the new algorithms instead of K-means to enhance the spectral clustering algorithm [174]. Figure 4.13 shows the results after applying spectral clustering algorithm with K-means, and by using different prototypes' initializations, to the artificial data set shown in Figure 4.12. As shown in Figure 4.13, the spectral clustering algorithm with

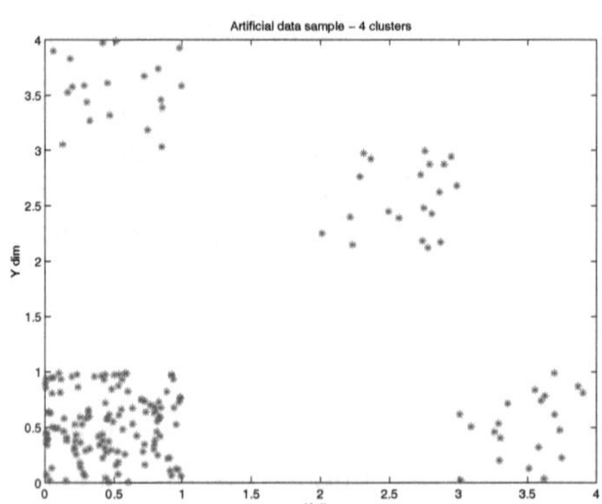

Fig. 4.12. Artificial data set consisting of 4 clusters of red '*'s

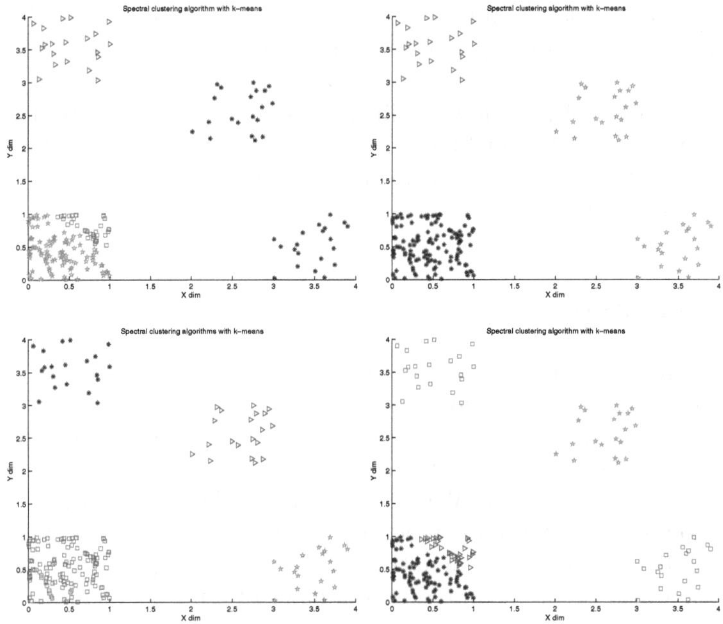

Fig. 4.13. Results of spectral clustering algorithm with K-means using different prototypes' initializations

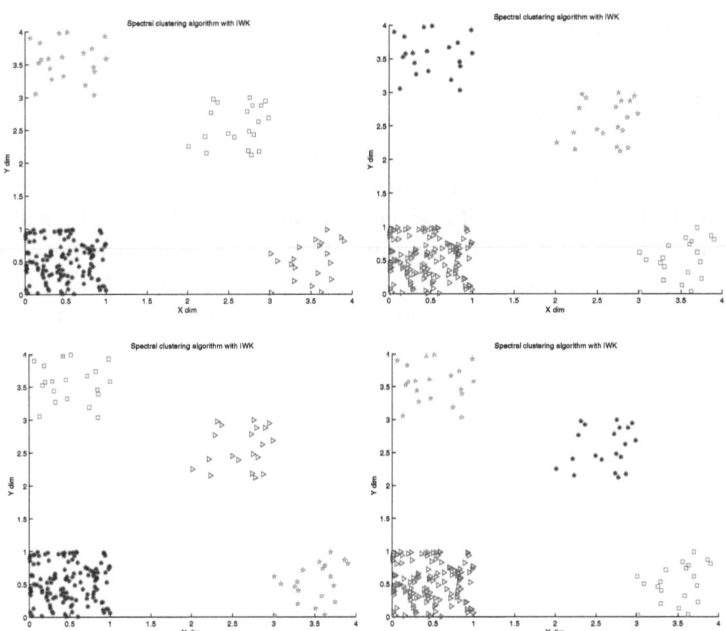

Fig. 4.14. Results of spectral clustering algorithm with IWK using different prototypes' initializations

K-means failed to identify all the clusters successfully and the algorithm still exhibits sensitivity to the protoypes' initializations. Figure 4.14 shows the results after applying spectral clustering algorithm with inverse weighted K-means algorithm (IWK), and with different prototypes' initializations, to the artificial data set shown in Figure 4.12. As shown in Figure 4.14, the new spectral clustering algorithm always gives better results and identifies all the clusters successfully.

4.3 Conclusion

We have investigated methods for alleviating one of the problems with K-means, a standard method for clustering data. This problem is based on its sensitivity to initial conditions, as a result of which it will often cause convergence of the prototypes to positions which are only locally optimal rather than globally optimal. We have shown how the incorporation of global information into a performance function enables us to create adaptation rules which find better prototypes than the simple K-means algorithm. The global information may be incorporated in a number of different manners, each of which has a slightly different rationale: the first simply enables us to minimise the total distance between all the prototypes and the data while the second moves the closest prototype to the current data point while pushing other prototypes away from the closest prototype.

The question arises as to which of the algorithms in this chapter is best: this is a question for which we have no set answer. The algorithm to use is very much a "horses for courses" decision: some algorithms seem to work best with some data sets while others are optimal for different data sets. We have found no overarching reasons as to why this should be. Therefore our advice to anyone involved in clustering a data set is to use as many different clustering techniques as possible since each may give different views of a data set and each view may contain valuable information about the datat set.

In the next chapter, we use these clustering algorithms as the basis of some topology preserving algorithms and for clustering in kernel spaces.

5

Topographic Mappings and Kernel Clustering

5.1 A Topology Preserving Mapping

A topographic mapping (or topology preserving mapping) is a transformation which captures some structure in the data so that points which are mapped close to one another share some common feature while points which are mapped far from one another do not share this feature. The Self-organizing Map (SOM) [137, 138] was introduced as a data quantisation method but has found at least as much use as a visualisation tool.

Topology-preserving mappings such as the Self-organizing Map (SOM) [138] and the Generative Topographic Mapping (GTM) [34] have been very popular for data visualization: we project the data onto the map which is usually two dimensional and look for structure in the projected map by eye. There has recently been proposed a family of topology preserving mappings [95] which are based on the same underlying structure as the GTM.

In this chapter we show how it is possible to extend the previous clustering algorithms to provide new algorithms for visualization and topology-preserving mappings. All the following algorithms have the same structure as the GTM, with a number of latent points that are mapped to a feature space by M Gaussian functions, and then into the data space by a matrix W. Each latent point $\mathbf{t}$ indexed by k is mapped, through a set of M fixed basis functions $\Phi_1(), \Phi_2(),...,\Phi_M()$ to a prototype in data space $\mathbf{m}_k = \mathbf{W}\Phi(\mathbf{t}_k)$.

1. Inverse-weighted K-means Topology-preserving Maping (IKToM):
   To construct IKToM and then visualize data, the mapped prototypes in data space are allocated by using Inverse Weighted K-means (IWK) algorithm.
2. Inverse Exponential K-means Topology-preserving Mapping 1 (IEK-ToM1):
   In IEKToM1 algorithm we use the Inverse Exponential K-means1 (IEK1) algorithm in learning process to train the prototypes appropriately. Finally, after convergence, we project the data samples onto a two dimensional grid of latent points using the trained prototypes.

W.A. Barbakh et al.: Non-Stand. Param. Adapt. for Explor. Data Analys., SCI 249, pp. 73–84.
springerlink.com

3. Inverse Exponential K-means Topology-preserving Mapping 2 (IEKToM2):
   IEKToM2 is similar to IEKToM1 but with using IEK2 algorithm instead of IEK1 in the learning process.
4. Inverse weighted Clustering Topology-preserving Mapping (ICToM): Similarly, in the ICToM algorithm we use the inverse weighted clustering algorithm (IWC) in the learning process under the GTM structure.

5.1.1 Simulations

In this section we apply the new topology preserving mapping algorithms to artificial and real data sets.

Artificial data set

We create a simulation with 20 latent points deemed to be equally spaced in a one dimensional latent space, passed through 5 Gaussian basis functions and then mapped to the data space by the linear mapping W which is the only parameter we adjust. We generated 500 two dimensional data points from the function $x_2 = x_1 + 1.25\sin(x_1) + \mu$ where μ is noise from a uniform distribution in [0,1]. Final results from the IKToM, IEKToM1, IEKToM2 and ICToM are shown in Figure 5.1. From Figure 5.1, we can see the prototypes fit the manifold, preserving the topology, in a good way for the new algorithms. Also, IEKToM1 and ICToM give a model that is more smooth than others. IEKToM2, being more data-hungry, gives the best results in identifying the manifold's ends.

Iris data set

In the iris data set we have 150 samples with 4 dimensions and 3 types.

Figure 5.2 shows the result after applying IKToM, IEKToM1, IEKToM2 and ICToM to the iris data set. The iris data set is projected onto a two dimensional grid of latent points (10 x 10). Most researchers will recognize the projection of the iris data set as familiar in that one class is readily separated while the other two are rather more difficult to separate.

Algae data set

In algae[1] data set we have 72 samples classified into 9 types. Each sample is recorded as an 18 dimensional vector representing the magnitudes of various pigments. Figure 5.3 shows the result after applying IKToM, IEKToM1, IEKToM2 and ICToM to the algae data set. The algae data set is projected onto a two dimensional grid of latent points (10 x 10). In each case, we see

[1] The algae data set is available at `http://mlearn.ics.uci.edu/databases/`.

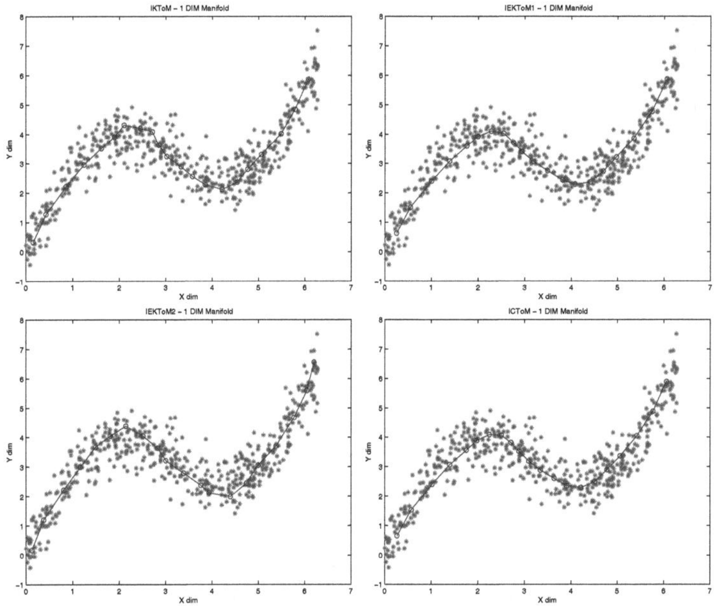

Fig. 5.1. The resulting prototypes' positions after applying IKToM, top left, IEKToM1, top right, IEKToM2, bottom left, and ICToM, bottom right. Prototypes are shown as blue 'o's.

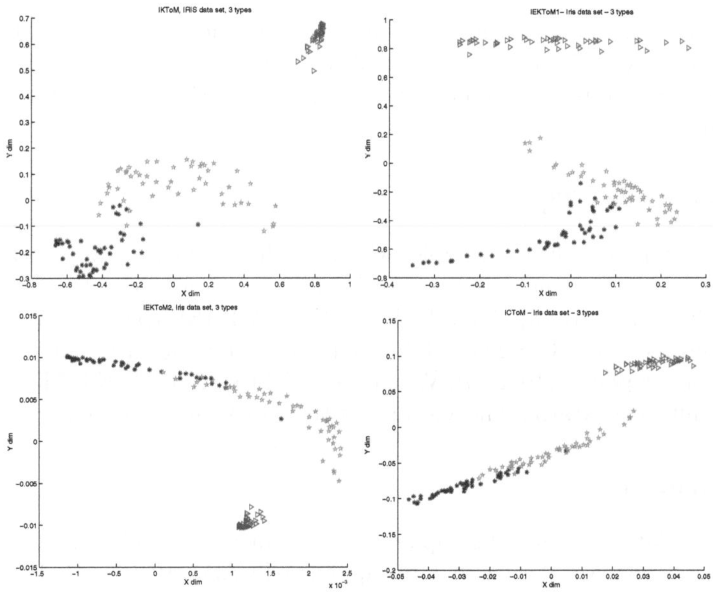

Fig. 5.2. Top left: the result of using IKToM on the iris data set. The top right:the result of using IEKToM1. Bottom left: the result of using IEKToM2. Bottom right: the result of using ICToM.

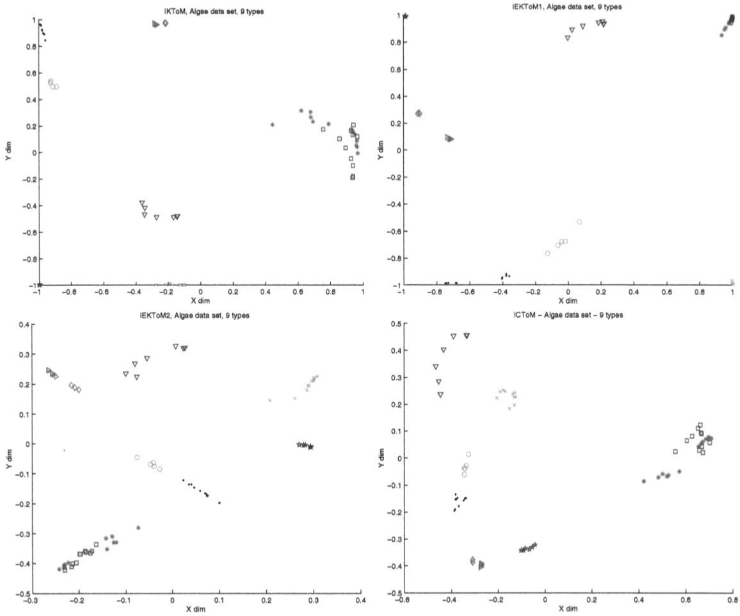

Fig. 5.3. Top left: the result of using IKToM on the algae data set. The top right:the result of using IEKToM1. Bottom left: the result of using IEKToM2. Bottom right: the result of using ICToM.

a projection of the classes into the latent space which gives a separation between the classes. 7 clusters are identified successfully. We still see that it is difficult to separate the two clusters represented by the blue squares and the red stars.

Genes data set [73]

The genes[2] data set has 40 samples with 3036 dimensions and 3 types of bladder cancer, i.e. extremely high dimensional and with few samples. Figure 5.4 shows the result after applying IKToM, IEKToM1, IEKToM2 and ICToM to the genes data set. The genes data set is projected onto a two dimensional grid of latent points (10 x 10). We can see IEKToM1 and IEKToM2 give the best result in separating and visualizing the three clusters of genes.

Glass data set

The glass data set has 214 samples with 10 dimensions and 6 types.

Figure 5.5 shows the result after applying IKToM, IEKToM1, IEKToM2 and ICToM to the Glass data set. The glass data set is projected onto a two

[2] The genes data set is available at `http://www.ihes.fr/~zinovyev/princmanif2006/`.

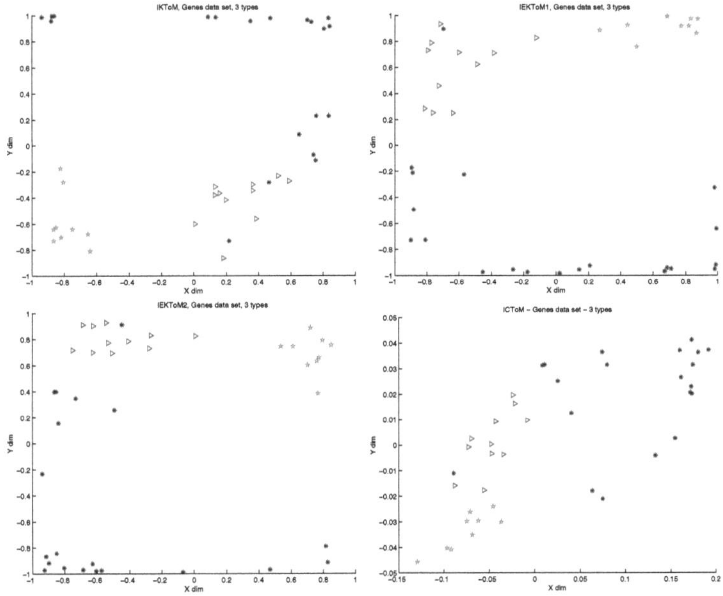

Fig. 5.4. Top left: the result of using IKToM on the genes data set. The top right:the result of using IEKToM1. Bottom left: the result of using IEKToM2. Bottom right: the result of using ICToM.

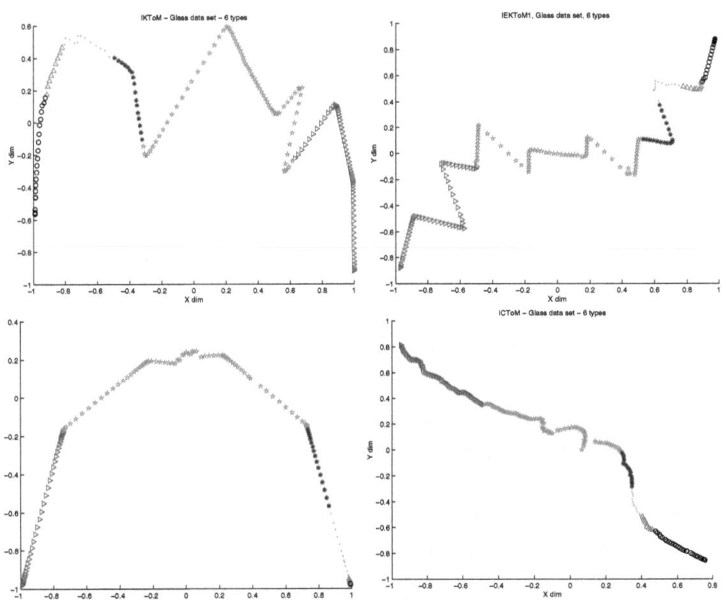

Fig. 5.5. Top left: the result of using IKToM on the glass data set. The top right:the result of using IEKToM1. Bottom left: the result of using IEKToM2. Bottom right: the result of using ICToM.

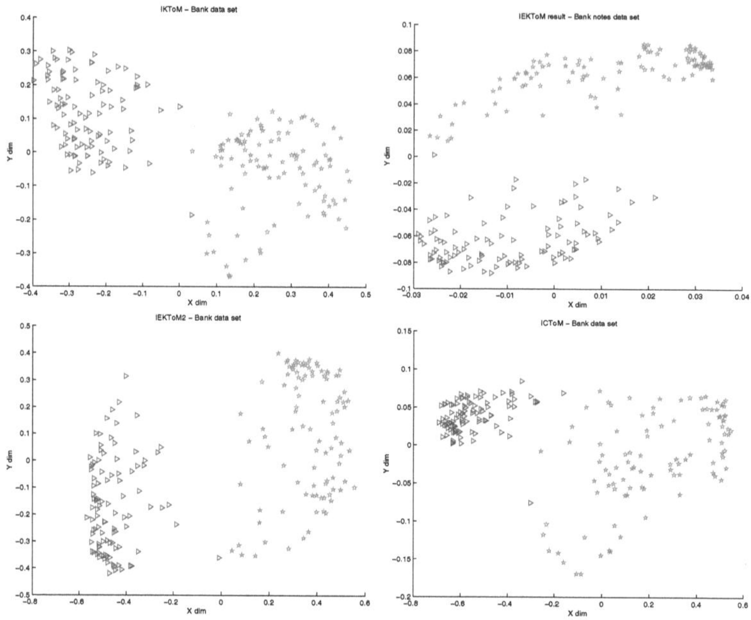

Fig. 5.6. Top left: the result of using IKToM on the bank data set. The top right:the result of using IEKToM1. Bottom left: the result of using IEKToM2. Bottom right: the result of using ICToM.

dimensional grid of latent points (10 x 10). IKToM, IEKToM1, and ICToM give similar results and show the 6 clusters as connected segments, while IEKToM2 gives the worst result as it combines the two clusters represented by black circles and pink triangles.

Bank data set

The Bank[3] data set appeared in [82]. This data set has 200 observations. It contains measurements on 100 forged and 100 genuine bank notes. Each data record contains the six measurements:

LENGTH: length of bill.
LEFT: width of bill, measured on the left.
RIGHT: width of bill, measured on the right.
BOTTOM: width of the margin at the bottom.
TOP: width of the margin at the top.
DIAGONAL: length of the image diagonal.

Figure 5.6 shows the result after applying IKToM, IEKToM1, IEKToM2 and ICToM to the Bank data set. The bank data set is projected onto a two

[3] The Bank data set is available at `http://www.quantlet.com/mdstat/scripts/mva/htmlbook/mvahtmlnode129.html`.

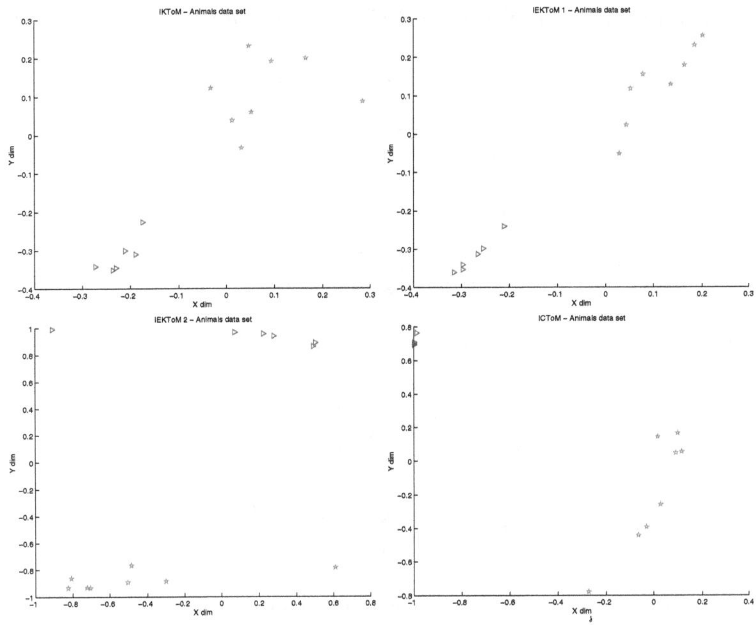

Fig. 5.7. Top left: the result of using IKToM on the animals data set. The top right:the result of using IEKToM1. Bottom left: the result of using IEKToM2. Bottom right: the result of using ICToM.

dimensional grid of latent points (10 x 10). IEKToM1 and IEKToM2 give the best result in separating and visualizing the two clusters of bank data.

Animals data set

The animals[4] data set has 16 animals which are described by 13 attributes such as size, legs, running, swimming or flying animals etc. These 16 animals are classified into two clusters, birds and mammals. This data set contains information on 7 birds and 9 mammals. Figure 5.7 shows the result after applying IKToM, IEKToM1, IEKToM2 and ICToM to the animals data set. The animals data set is projected onto a two dimensional grid of latent points (10 x 10). All the new algorithms succeed in identifying and visualizing the animals data set.

5.2 Kernel Clustering Algorithms

In this section, we show how to extend some of our algorithms in Chapter 3 to construct new kernel clustering algorithms.

[4] The animals data set is available at `http://www.ifs.tuwien.ac.at/ifs/research/pub_html/rau_ijcnn99/ijcnn99.html`.

5.2.1 Kernel Inverse Weighted Clustering Algorithm (KIWC)

In the inverse weighted clustering algorithm (IWC) we have:

$$\mathbf{m}_k = \frac{\sum_{i=1}^{N} b_{ik}\mathbf{x}_i}{\sum_{i=1}^{N} b_{ik}} \tag{5.1}$$

where

$$b_{ik} = \frac{\| \mathbf{x}_i - \mathbf{m}_{k*} \|^{P+2}}{\| \mathbf{x}_i - \mathbf{m}_k \|^{P+2}} \tag{5.2}$$

For kernel space we can rewrite (5.1) as follows:

$$\mathbf{m}_k = \frac{\sum_{i=1}^{N} b'_{ik}\phi(\mathbf{x}_i)}{\sum_{i=1}^{N} b'_{ik}} \tag{5.3}$$

where

$$b'_{ik} = \frac{\| \phi(\mathbf{x}_i) - \mathbf{m}_{z*} \|^{P+2}}{\| \phi(\mathbf{x}_i) - \mathbf{m}_k \|^{P+2}} \tag{5.4}$$

and $\mathbf{m}_{z*}$ represents the closest prototype in the kernel space to $\phi(\mathbf{x}_i)$. Note that

$$\begin{aligned} \| \phi(\mathbf{x}_i) - \mathbf{m}_k \|^2 &= \| \phi(\mathbf{x}_i) - \frac{\sum_{s=1}^{N} b'_{sk}\phi(\mathbf{x}_i)}{\sum_{s=1}^{N} b'_{sk}} \|^2 \\ &= kernel(\mathbf{x}_i, \mathbf{x}_i) - \frac{2}{\sum_{s=1}^{N} b'_{sk}} \left(\sum_{s=1}^{N} b'_{sk} kernel(\mathbf{x}_i, \mathbf{x}_s) \right) \\ &+ \frac{1}{(\sum_{s=1}^{N} b'_{sk})^2} \left(\sum_{s=1}^{N} \sum_{l=1}^{N} b'_{lk} b'_{sk} kernel(\mathbf{x}_s, \mathbf{x}_l) \right) \end{aligned} \tag{5.5}$$

The kernel inverse weighted clustering algorithm can be implemented through the following steps:

1. Initialize the b'_{ik} randomly for all pairs of data points and prototypes.
2. Construct the kernel matrix for all data points.
3. Calculate the distances, in kernel space, between all data points and prototypes using (5.5).
4. Calculate b'_{ik} using (5.4).
5. Repeat steps 3 and 4 until convergence.
6. Assign every data point $\mathbf{x}_i$ to the prototype $\mathbf{m}_k$ that gives the maximum b'_{ik}.

5.2.2 Kernel K-Harmonic Means Algorithm (KKHM)

For kernel K-Harmonic means algorithm we use the same formulation.

$$\mathbf{m}_k = \frac{\sum_{i=1}^{N} b'_{ik}\phi(\mathbf{x}_i)}{\sum_{i=1}^{N} b'_{ik}} \tag{5.6}$$

where

$$b'_{ik} = \frac{1}{\| \phi(\mathbf{x}_i) - \mathbf{m}_k \|^4 \left(\sum_{l=1}^{K} \frac{1}{\|\phi(\mathbf{x}_i)-\mathbf{m}_l\|^2}\right)^2} \tag{5.7}$$

The implementation of this algorithm is similar to the implementation of (KIWC) algorithm, we use (5.7) instead of (5.4). We have got similar results as in (KIWC) algorithm.

5.2.3 Kernel Inverse Weighted K-Means Algorithm (KIWK)

In inverse weighted K-means algorithm we have two sets of updates, see (4.18), (4.19) and (4.20). For kernel space we can rewrite them as follows:

$$\mathbf{m}_r = \frac{\sum_{i\in V_r} \phi(\mathbf{x}_i)a'_{ir} + \sum_{i\in V_j, j\neq r} \phi(\mathbf{x}_i)b'_{ir}}{\sum_{i\in V_r} a'_{ir} + \sum_{i\in V_j, j\neq r} b'_{ir}} \tag{5.8}$$

where V_r contains the indices of data points, in kernel space, that are closest to $\mathbf{m}_r$, V_j contains the indices of all the other points and

$$\begin{aligned} a'_{ir} &= -(n-p) \| \phi(\mathbf{x}_i) - \mathbf{m}_r \|^{n-p-2} \\ &\quad -n \| \phi(\mathbf{x}_i) - \mathbf{m}_r \|^{n-2} \sum_{j\neq z*} \frac{1}{\| \phi(\mathbf{x}_i) - \mathbf{m}_j \|^p} \end{aligned} \tag{5.9}$$

$$b'_{ir} = p\frac{\| \phi(\mathbf{x}_i) - \mathbf{m}_{z*} \|^n}{\| \phi(\mathbf{x}_i) - \mathbf{m}_r \|^{p+2}} \tag{5.10}$$

where $\mathbf{m}_{z*}$ is the closest prototype in the kernel space to $\phi(\mathbf{x}_i)$.

The implementation of this algorithm is similar to the implementation of (KIWC) algorithm, although we have here two sets of updates instead of one.

5.2.4 Simulations

We applied kernel K-means algorithm, kernel inverse weighted clustering (KIWC) algorithm and kernel inverse weighted K-means (KIWK) to the artificial data set shown in Figure 5.8. Figure 5.9 shows the results after applying kernel K-means, and with using different prototypes' initializations. While kernel K-means gave different results and failed with some initializations to identify the clusters successfully, the new algorithms (KIWC), Figure 5.10, and (KIWK), Figure 5.11, identified all the clusters successfully even if we have different or poor prototypes' intializations.

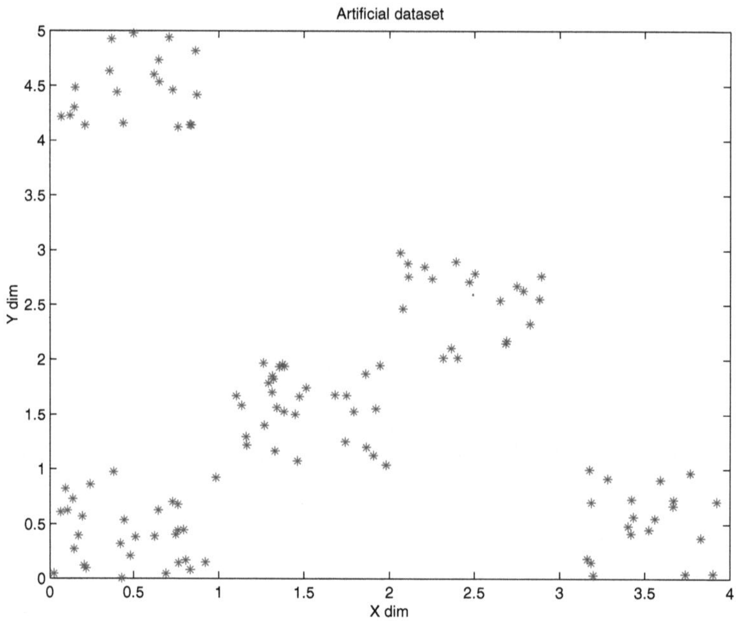

Fig. 5.8. An artificial data set consisting of 5 clusters

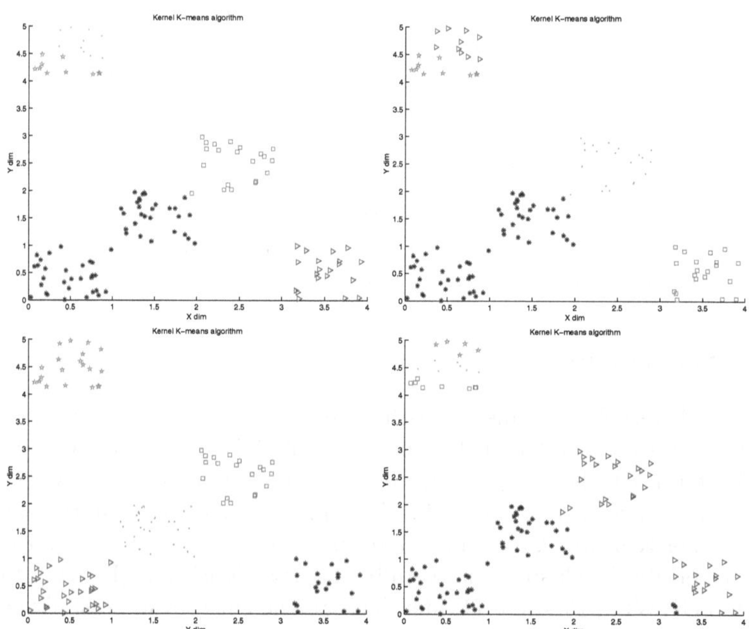

Fig. 5.9. The results of applying kernel K-means algorithm to the dataset in Figure 5.8

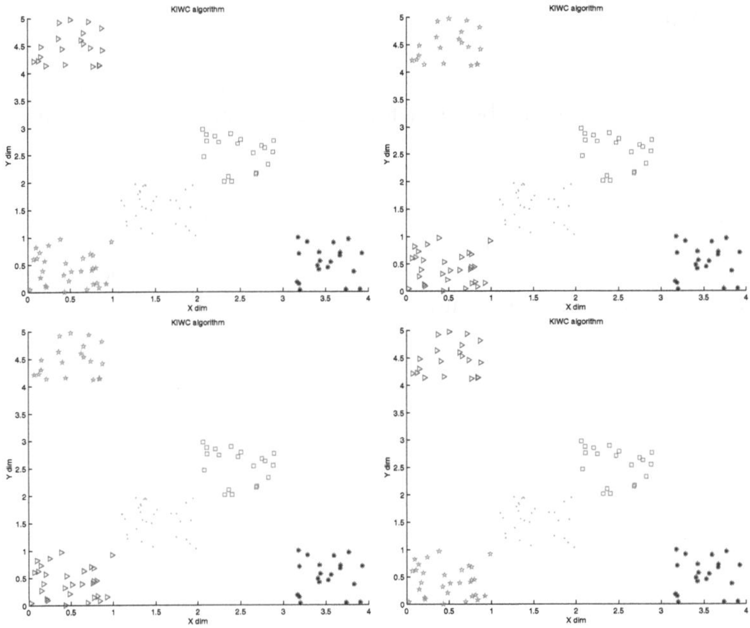

Fig. 5.10. The results of applying KIWC algorithm to the dataset in Figure 5.8

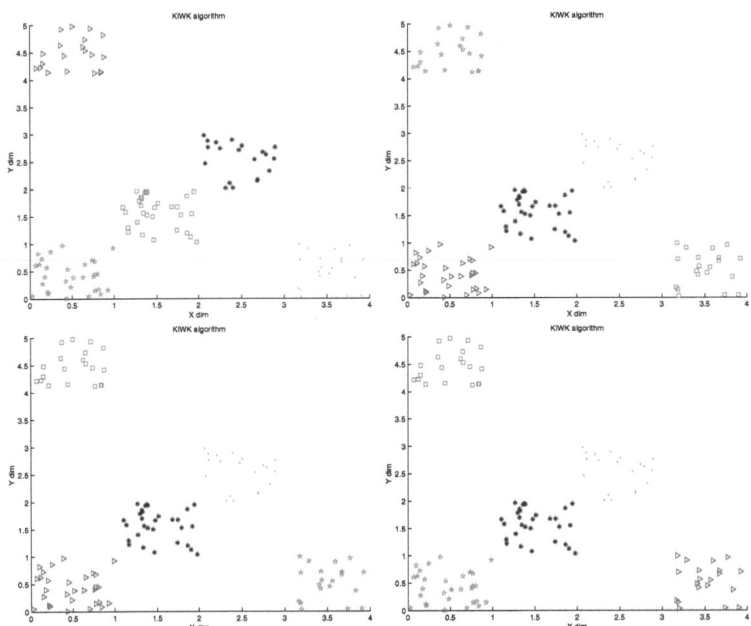

Fig. 5.11. The results of applying KIWK algorithm to the dataset in Figure 5.8

5.3 Conclusion

We have shown how we can extend any one of the new algorithms presented in the previous chapter to be used for visualization and dimensionality reduction.

We created a latent space containing latent points which are mapped to the prototypes and showed how these new rules may be used to create topology-preserving maps. We have illustrated such maps on artificial and real data sets. Finally, we have shown how to extend any one of the previous algorithms to construct new kernel clustering algorithms.

6

Online Clustering Algorithms and Reinforcement Learning

6.1 Online Clustering Algorithms

In this chapter, we show how we can extend the algorithms in Chapter 3 and allow them to learn in online mode. The aim of this chapter is to allow prototypes to learn in a different way, online, to that in batch mode. This may lead to different results due to the different behavior in the learning process. Furthermore, a limitation of batch processing algorithms is that they cannot readily respond to new data if the data only becomes available over time. Thus we construct a new set of online clustering algorithms based on extension of some of the algorithms in Chapter 3 and sharing the same performance functions.

6.1.1 Online K-Means Algorithm

The performance function for K-means may be written as

$$J_{Km} = \sum_{i=1}^{N} \min_{j=1}^{K} \| \mathbf{x}_i - \mathbf{m}_j \|^2 \tag{6.1}$$

The implementation of the online K-means algorithm is as follows [247]:

1. Initialization: initialize the cluster prototype vectors $\mathbf{m}_1, ..., \mathbf{m}_K$
2. Loop for M iterations:
   a) for each data vector $\mathbf{x}_i$, set

$$k^* = \arg\min_{k=1}^{K}(\| \mathbf{x}_i - \mathbf{m}_k \|) \tag{6.2}$$

   b) update the prototype $\mathbf{m}_{k^*}$ as

$$\mathbf{m}_{k^*}^{(new)} = \mathbf{m}_{k^*} - \zeta \frac{\partial J_1}{\partial \mathbf{m}_{k^*}} = \mathbf{m}_{k^*} + \zeta(\mathbf{x}_i - \mathbf{m}_{k^*}) \tag{6.3}$$

   where ζ is a learning rate usually set to be a small positive number.

The learning rate can also gradually decreases during the learning process.

W.A. Barbakh et al.: Non-Stand. Param. Adapt. for Explor. Data Analys., SCI 249, pp. 85–108.
springerlink.com

6.1.2 IWK Online Algorithm v1 (IWKO1)

We note that some researchers differentiate between online and incremental algorithms:

Online: This word is reserved for algorithms used when the data is available only one piece at a time whereas

Incemental: can be used for algorithms which slowly change the parameters regardless of when the data becomes available.

We do not make this distinction in this book and use the words interchangeably.

From Chapter 3, for IWK (batch mode) we have the following:

$$J_{IWK} = \sum_{i=1}^{N} \left[\sum_{j=1}^{K} \frac{1}{\| \mathbf{x}_i - \mathbf{m}_j \|^p} \right] \min_{k=1}^{K} \| \mathbf{x}_i - \mathbf{m}_k \|^n . \tag{6.4}$$

$$\mathbf{m}_k = \frac{a_1 \mathbf{x}_1 + ... + a_N \mathbf{x}_N}{a_1 + ... + a_N} \tag{6.5}$$

where

$$a_i = \begin{cases} \left(-(n-p) \| \mathbf{x}_i - \mathbf{m}_k \|^{n-p-2} - n \| \mathbf{x}_i - \mathbf{m}_k \|^{n-2} \sum_{j \neq k*} \frac{1}{\|\mathbf{x}_i - \mathbf{m}_j\|^p} \right) & k = k^* \\ \left(p \frac{\|\mathbf{x}_i - \mathbf{m}_{k*}\|^n}{\|\mathbf{x}_i - \mathbf{m}_k\|^{p+2}} \right) & \text{otherwise} \end{cases}$$

and $\mathbf{m}_{k^*}$ is the closest prototype to $\mathbf{x}_i$.

Implementation (Online Mode)

In online mode, for IWK we can do something similar to (6.5) (taking into account that we receive one input sample at a time) as follows:

1. Initialization:
   - initialize the cluster prototype vectors $\mathbf{m}_1, ..., \mathbf{m}_K$
   - initialize one dimensional vector, v, with K elements to one, $v_1 = 1, ..., v_K = 1$
   note: v_k will represent the value that should be in the denominator of (6.5) after feeding the input sample (it is used also for normalization).
2. Loop for M iterations:
   - for each data vector $\mathbf{x}_i$, set

$$k^* = \arg \min_{k=1}^{K} (\| \mathbf{x}_i - \mathbf{m}_k \|) \tag{6.6}$$

and update all the prototypes $\mathbf{m}_k$ as

$$\mathbf{m}_k^{(new)} = \frac{\mathbf{m}_k v_k + a_i \mathbf{x}_i}{v_k + a_i}$$
$$v_k{}^{(new)} = v_k + a_i$$

To clarify how this implementation is going to build (6.5), we will feed two data points as an example. After feeding the first data point $\mathbf{x}_1$, we will have the following:

$$\mathbf{m}_k^{(new)} = \frac{(\mathbf{m}_k * 1) + a_1 \mathbf{x}_1}{1 + a_1}$$
$$v_k{}^{(new)} = 1 + a_1$$

After feeding the second data point $\mathbf{x}_2$, we will have the following:

$$\mathbf{m}_k^{(new)} = \frac{\left(\frac{(\mathbf{m}_k * 1) + a_1 \mathbf{x}_1}{1 + a_1}\right)(1 + a_1) + a_2 \mathbf{x}_2}{1 + a_1 + a_2}$$
$$= \frac{\mathbf{m}_k + a_1 \mathbf{x}_1 + a_2 \mathbf{x}_2}{1 + a_1 + a_2}$$
$$v_k{}^{(new)} = 1 + a_1 + a_2$$

And so on in the same way for all data points. In this version we try to implement the batch mode algorithm but when we have the data one at a time. May we need before the end of the execution to subtract $\mathbf{m}_k$ from the numerator and 1 from the denominator to have (6.5).

Simulations

Figure 6.1, right, shows the results after applying IWKO1 to the artificial data set in Figure 6.1, left.

6.1.3 IWK Online Algorithm v2 (IWKO2)

In this section we show how it is possible to allow all the units (prototypes) to learn, not only the winner as in K-means or the winner with its neighbors as in SOM. In this algorithm, it is not necessary to specify any functions for the neighbors as all units learn with every input sample.

Implementation (Online Mode)

1. Initialization:
   - initialize the cluster prototype vectors $\mathbf{m}_1, ..., \mathbf{m}_K$

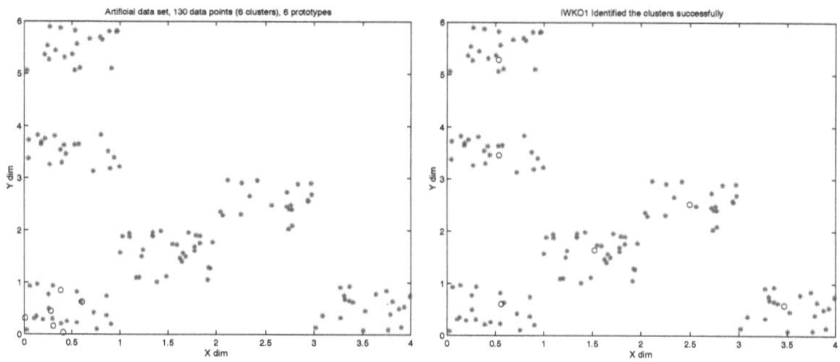

Fig. 6.1. Left: Artificial data set: data set is shown as 6 clusters of red '*'s, prototypes are initialized to lie within one cluster and shown as blue 'o's. Right: IWKO1 succeeds in identifying all the clusters.

2. Loop for M iterations:
   - from (4.15) and (4.16) and ($p = -1$) we have:

$$\Delta \mathbf{m}_{k^*} = -\zeta(\mathbf{x}_i - \mathbf{m}_{k^*})a_{ik^*} \tag{6.7}$$
$$\Delta \mathbf{m}_k = -\zeta(\mathbf{x}_i - \mathbf{m}_k)b_{ik} \tag{6.8}$$

   - for each data vector $\mathbf{x}_i$, set

$$k^* = \arg\min_{k=1}^{K}(\| \mathbf{x}_i - \mathbf{m}_k \|) \tag{6.9}$$

   and update all the prototypes $\mathbf{m}_k$ as

$$\mathbf{m}_{k^*}^{(new)} = \mathbf{m}_{k^*} - \zeta a_{ik^*}(\mathbf{x}_i - \mathbf{m}_{k^*})$$

   where

$$a_{ik^*} = -(n+1) \| \mathbf{x}_i - \mathbf{m}_{k^*} \|^{n-1} - n \| \mathbf{x}_i - \mathbf{m}_{k^*} \|^{n-2} \sum_{j \neq k^*} \| \mathbf{x}_i - \mathbf{m}_j \|$$

$$\mathbf{m}_k^{(new)} = \mathbf{m}_k - \zeta \left(\frac{- \| \mathbf{x}_i - \mathbf{m}_{k^*} \|^n}{\| \mathbf{x}_i - \mathbf{m}_k \|} \right) (\mathbf{x}_i - \mathbf{m}_k)$$

   where ζ is a learning rate.

Note: If we want to choose a bigger value for n, we will need to choose smaller value for ζ (thus we tend to choose $n = 1$).

Simulations

We applied IWKO2 and K-means to some artificial data sets. We found K-means failed to identify all the clusters due to the bad initialization while IWKO2 algorithm succeeded. Figure 6.3 shows the results after applying K-means, left, and IWKO2, right, to the artificial data set shown in Figure 6.2.

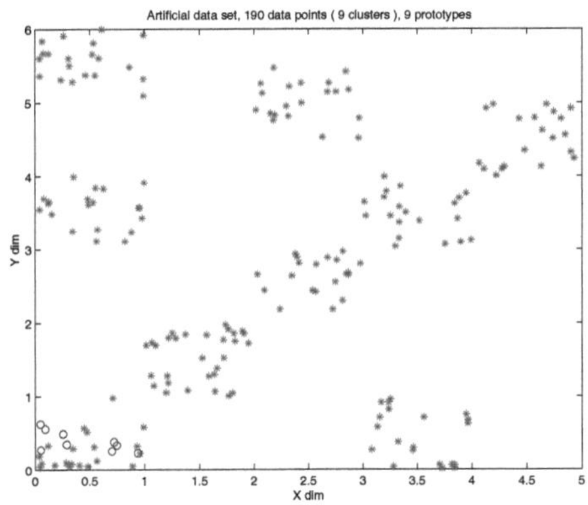

Fig. 6.2. Artificial data set: data set is shown as 6 clusters of red '*'s, prototypes are initialized to lie within one cluster and shown as blue 'o's.

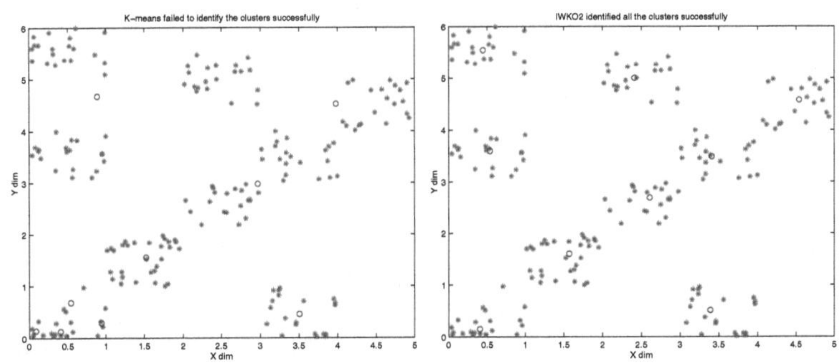

Fig. 6.3. Left: K-means results. Right: IWKO2 succeeds in identifying all the clusters.

Note that K-means can succeed with this artificial data set by choosing good prototypes' initialization, but we want to show that the new algorithm IWKO2 still works well even if we have a bad initialization in which K-means failed.

Note that for this algorithm, it is also possible to work in the same manner as the SOM and update the winner with its neighbors instead of updating all the units. In this way, we could create an algorithm that has the advantages of both SOM and IWKO2.

6.1.4 K-Harmonic Means – Online Mode Algorithm (KHMO)

In K-Harmonic means we have the following performance function:

$$J_{HA} = \sum_{i=1}^{N} \frac{K}{\sum_{k=1}^{K} \frac{1}{\|\mathbf{x}_i - \mathbf{m}_k\|^2}}. \tag{6.10}$$

This was used in batch mode in [241, 244]. We can also create an online version of K-Harmonic means algorithm.

Implementation (Online Mode)

From (2.15) we have:

$$\frac{\partial J_{HA}(\mathbf{x}_i)}{\partial \mathbf{m}_k} = -K \frac{2(\mathbf{x}_i - \mathbf{m}_k)}{\| \mathbf{x}_i - \mathbf{m}_k \|^4 \{\sum_{l=1}^{K} \frac{1}{\|\mathbf{x}_i - \mathbf{m}_l\|^2}\}^2} \tag{6.11}$$

$$\begin{aligned} \mathbf{m}_k^{(new)} &= \mathbf{m}_k + \zeta \left(\frac{2K}{\| \mathbf{x}_i - \mathbf{m}_k \|^4 \{\sum_{l=1}^{K} \frac{1}{\|\mathbf{x}_i - \mathbf{m}_l\|^2}\}^2} \right) (\mathbf{x}_i - \mathbf{m}_k) \\ &= \mathbf{m}_k + \zeta \left(\frac{1}{\| \mathbf{x}_i - \mathbf{m}_k \|^4 \{\sum_{l=1}^{K} \frac{1}{\|\mathbf{x}_i - \mathbf{m}_l\|^2}\}^2} \right) (\mathbf{x}_i - \mathbf{m}_k) \end{aligned}$$

where ζ is a learning rate usually set to be a small positive number.

Simulations

In Figure 6.4, left, we have an artificial data set consisting of 9 clusters. All the prototypes are initialized poorly within the first cluster. Figure 6.4, right, shows the results after applying KHMO to this artificial data set. KHMO succeeds in identifying all the clusters successfully even though we have a poor initialization.

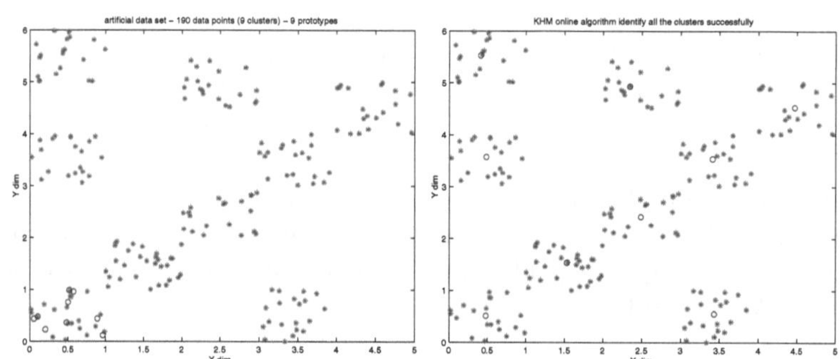

Fig. 6.4. Left: artificial data set: the data set is shown as 9 clusters of red '*'s; the prototypes are initialized to lie within one cluster and shown as blue 'o's. Right: KHMO succeeds in identifying all the clusters.

6.1.5 Inverse-Weighted K-Means (Online) Topology-Preserving Mapping (IKoToM)

IKoToM algorithm has the same structure as the GTM, with a number of latent points that are mapped to a feature space by M Gaussian functions, and then into the data space by a matrix W. The IKoToM uses the inverse weighted online clustering algorithm for the learning process.

Simulations

Artificial data set

We create a simulation with 10 latent points deemed to be equally spaced in a one dimensional latent space, passed through 5 Gaussian basis functions and then mapped to the data space by the linear mapping W which is the only parameter we adjust. We generated 500 two dimensional data points, (x_1, x_2), from the function $x_2 = x_1 + 1.25\sin(x_1) + \mu$ where μ is noise from a uniform distribution in [0,1]. Final result from the IKoToM is shown in Figure 6.5 in which the projections of consecutive latent points are joined. We see that nearby latent points take responsibility for nearby data points.

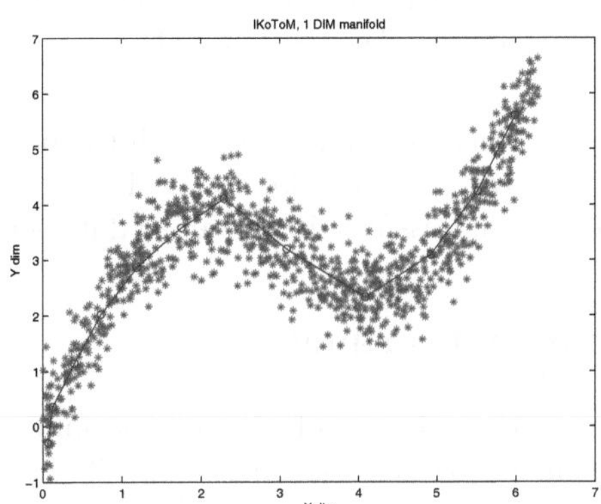

Fig. 6.5. The resulting prototypes' positions after applying IKoToM. The prototypes are shown as blue 'o's.

Real data sets

We show in Figure 6.6 the projections of the real data sets iris, algae, genes and glass onto a two dimensional grid of latent points (10 x 10) using IKoToM. In each one, we see a projection of the classes into the latent space which gives a separation between the classes: most researchers will recognise the projection of the iris data set as familiar in that one class is readily separated while the other two are rather more difficult to separate; on the other data

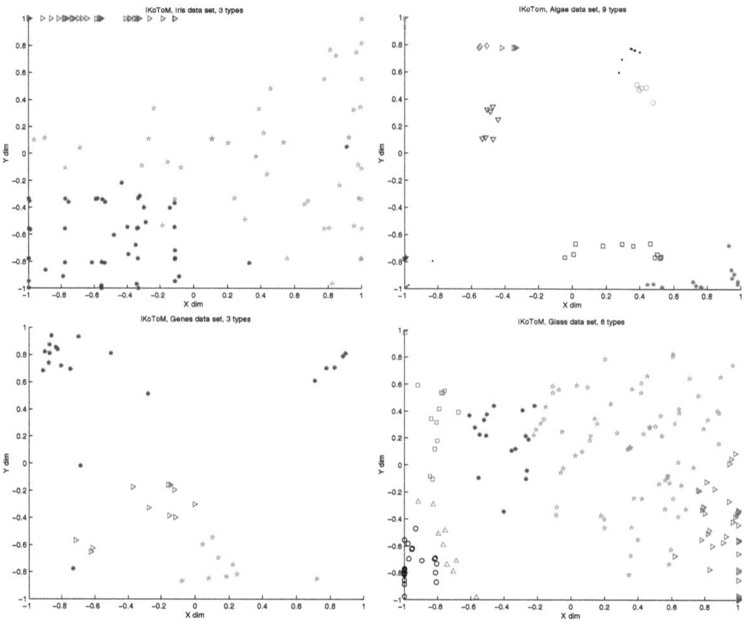

Fig. 6.6. The results of using IKoToM on the real data sets, iris, top left, algae, top right, genes, bottom left, and glass, bottom right.

sets, we achieve results which are comparable with the best projections found by other methods we have used [11, 99, 100]. Also IKoToM visulaizes and identifies all the clusters successfully in the algae data set.

6.2 Reinforcement Learning

Reinforcement learning is a sub-area of machine learning which trains agents by reward and punishment without needing to specify how to achieve a task. Reinforcement learning algorithms attempt to find a policy that maps states to actions so as to maximize some notion of long-term reward. It has been applied successfully to various problems, including robot control [160] and dispatch problems [53]. During the past two decades, the framework of reinforcement learning has encompassed a broad variety of problems such as clustering [150], classification [165] and function optimization [231].

The standard reinforcement learning model is described in [130], which consists of:

1. a set of environment states S;
2. a set of actions A;
3. a set of scalar "rewards" in $\Re$.

An agent exists within its environment and at each time of interaction, t, the agent perceives the state $s_t \in S$ of the environment and the set of possible actions $A(s_t)$. Then the agent chooses an action $a \in A(s_t)$ that changes the state of the environment from s_t to s_{t+1} and receives a reward r_{t+1}. The agent's behavior, B, should be based on a policy π, mapping states to actions, that tends to maximise the long-term sum of values of the rewards.

Reinforcement learning can therefore be thought of as an intermediate learning method that is distinct from supervised learning and unsupervised learning. There is no presentation of input-output pairs to the reinforcement learning machine, but the machine gets rewards from the external environment depending on how well it has done with an action and it must choose actions that tend to increase the long-term sum of values of rewards. We will in this book apply reinforcement learning methods to what are normally considered unsupervised learning problems.

There are two main strategies for solving reinforcement learning problems. The first is to search in the policy space directly by defining the policy as a parameterized function $\pi(s, \theta)$ with parameters θ, where commonly, a gradient method is employed to adjust the parameters. An alternative way is to use statistical techniques to estimate the utility of taking actions in states of the environment. Algorithms such as immediate reward reinforcement learning and evolutionary reinforcement learning are closer to the first strategy and algorithms such as temporal difference reinforcement learning tend to use the second strategy. In the following we give details of immediate reward reinforcement learning and temporal difference reinforcement learning. Evolutionary algorithms for reinforcement learning will also be very briefly introduced. This is done more for completeness than because the methods previewed here will be used in the remainder of the book.

6.2.1 Immediate Reward Reinforcement Learning

Immediate reward reinforcement learning assumes reward for an action is immediate, which means the learning is determined by the most recent state-action pair. In this method, we envisage stochastic learning units each of which has its own individual parameters; each unit at each instant in time may or may not fire and expectations in the following are with respect to the probability density functions associated with these stochastic units. A very natural performance measure to optimize for immediate reward reinforcement learning problems is the expected value of the reinforcement reward, r, conditioned on the parameters of the agents. Thus, by denoting $\mathbf{W}$ as the collection of parameters on which the behavior of agents depends, Williams [231, 230] considers the performance measure $E_{\mathbf{x}}\{r|\mathbf{W}\}$, where $E_{\mathbf{x}}\{r|\mathbf{W}\}$ is a well-defined, deterministic function of parameters $\mathbf{W}$. In this book, the expectations are always with respect to the data set and so we will in future omit the explicit subscript $\mathbf{x}$. Thus, the objective of immediate reward reinforcement learning is to search for a point where $E\{r|\mathbf{W}\}$ is maximum in the space of all possible parameters $\mathbf{W}$, so that a gradient-following algorithm is performed here.

Williams [230] related $\nabla_{\mathbf{W}} E\{r|\mathbf{W}\}$, the gradient in weight space of the performance measure $E\{r|\mathbf{W}\}$, to $E\{\Delta\mathbf{W}|\mathbf{W}\}$, the update vector in weight space, by introducing stochastic learning units in which the probability of output y_i was a function of the parameters of unit $\mathbf{w}_i$ and its input $\mathbf{x}$. For the ith unit, this gives

$$P(y_i = \xi|\mathbf{w}_i, \mathbf{x}) = f(\mathbf{w}_i, \mathbf{x}) \tag{6.12}$$

Typically we will be interested in Bernoulli units in which $\xi = 1$. The function $f()$ may be, for example,

$$f(\mathbf{w}_i, \mathbf{x}) = \frac{1}{1 + \exp(-||\mathbf{w}_i - \mathbf{x}||^2)} \tag{6.13}$$

The weights are adjusted following receipt of the reinforcement reward, r, at each iteration; each parameter w_{ij} is updated by

$$\Delta w_{ij} = \alpha_{ij}(r_{i,\xi} - b_{ij})\frac{\partial \ln P(y_i = \xi|\mathbf{w}_i, \mathbf{x})}{\partial w_{ij}} \tag{6.14}$$

where α_{ij} is a learning rate, b_{ij} is a reinforcement baseline (often $b_{ij} = \bar{r}_{ij}$, which will be used in this book) and $\partial \ln P(y_i = \xi|\mathbf{w}_i, \mathbf{x})/\partial w_{ij}$ is called the *characteristic eligibility* or *score function.* Williams [230] has proved that the average update vector in weight space corresponds to the direction for which the performance measure is increasing, so for each weight w_{ij}, the value of $(r_{i,\xi} - b_{ij})\partial \ln P(y_i = \xi|\mathbf{w}_i, \mathbf{x})/\partial w_{ij}$ represents an unbiased estimate of $\partial E\{r|\mathbf{W}\}/\partial w_{ij}$.

The way in which the unit may compute in [230] is to perform a deterministic computation based on its input and weights, and then draw its output randomly from an appropriate distribution. [230] gave the example of a Bernoulli unit in which $P(y_i = 1) = p_i$ and so $P(y_i = 0) = 1 - p_i$. Therefore

$$\frac{\partial \ln P(y_i)}{\partial p_i} = \begin{cases} -\frac{1}{1-p_i} & \text{if } y_i = 0 \\ \frac{1}{p_i} & \text{if } y_i = 1 \end{cases} = \frac{y_i - p_i}{p_i(1 - p_i)} \tag{6.15}$$

[150] applies the Bernoulli model to (unsupervised) clustering with

$$p_i = 2(1 - f(\mathbf{w}_i, \mathbf{x})) = 2\left(1 - \frac{1}{1 + \exp(- \parallel \mathbf{w}_i - \mathbf{x} \parallel^2)}\right) \tag{6.16}$$

The environment identifies the p_{i^*} which is maximum over all output units and y_{i^*} is then drawn from this distribution. Rewards are given such that

$$r_i = \begin{cases} 1 & \text{if } i = i^* \text{ and } y_i = 1 \\ -1 & \text{if } i = i^* \text{ and } y_i = 0 \\ 0 & \text{if } i \neq i^* \end{cases} \tag{6.17}$$

This is used in the update rule

$$\begin{aligned} \Delta w_{ij} &= \alpha r_i (y_i - p_i)(x_j - w_{ij}) && (6.18) \\ &= \alpha |y_i - p_i| (x_j - w_{ij}) \text{ for } i = i^* && (6.19) \end{aligned}$$

which is shown to perform clustering of the data set.

The stochastic units may also determine their output from multiparameter distributions in the same way. Assuming a Gaussian distribution is used here, the unit drawing its output according to such a distribution first computes values of mean μ and standard deviation σ deterministically and then determines its output from a normal distribution with μ and σ. Here, the μ and σ can be regarded as the adaptable parameters of the unit. Thus, the density function g determining the output y is given by

$$g(y, \mu, \sigma) = \frac{1}{(2\pi)^{1/2}\sigma} e^{-\frac{(y-\mu)^2}{2\sigma^2}} \tag{6.20}$$

The characteristic eligibility of μ is then

$$\frac{\partial \ln g}{\partial \mu} = \frac{(y - \mu)}{\sigma^2} \tag{6.21}$$

and the characteristic eligibility of σ is

$$\frac{\partial \ln g}{\partial \sigma} = \frac{(y - \mu)^2 - \sigma^2}{\sigma^3} \tag{6.22}$$

According to (6.14), the updates can be derived in the form of

$$\Delta \mu = \alpha_\mu (r - b) \frac{y - \mu}{\sigma^2} \tag{6.23}$$

and

$$\Delta \sigma = \alpha_\sigma (r - b) \frac{(y - \mu)^2 - \sigma^2}{\sigma^3} \tag{6.24}$$

Therefore, the weight is adjusted in the direction of the gradient of expected reward in immediate reward reinforcement tasks without specifying the explicit gradient estimates, $\nabla_{\mathbf{W}} E\{r|\mathbf{W}\}$.

6.2.2 Global Reinforcement Learning in Neural Networks with Stochastic Synapses

Recently, Ma and Likharev [155] have given an alternative view of supervised neural networks with reinforcement learning based on a slightly different formulation of the reinforcement learning principle described in the previous subsection: the networks are composed of deterministic nodes but stochastic synapses. The input signals $\mathbf{x}$ with zero mean are sent to outputs of other cells through synapses with weights w_{ij}:

$$y_i = \sum_j w_{ij} x_j. \tag{6.25}$$

In the first set of learning rules, [155] considers a neural network with deterministic cells but connected by a set of stochastic weights, w_{ij}, each of which is drawn from a Gaussian distribution $\mathcal{N}(\mu_{ij}, \sigma_{ij}^2)$ with mean μ_{ij} and variance, σ_{ij}^2. Then the output of this deterministic neuron, y_i, is also a Gaussian random variable $p_i(y_i) = Z \exp\left[-\frac{(y_i - \mu_{y_i})^2}{2\sigma_{y_i}^2}\right]$ with mean $\mu_{y_i} = \sum_j \mu_{ij} x_j$ and variance $\sigma_{y_i}^2 = \sum_j \sigma_{ij}^2 x_j^2$, where Z is a normalization term. Then we have

$$\frac{\partial \ln p_i(y_i)}{\partial \mu_{ij}} = \frac{(y_i - \mu_{y_i}) x_j}{\sigma_{y_i}^2} \tag{6.26}$$

According to (6.14), the learning rule is defined by

$$\Delta \mu_{ij} = \eta r \frac{(y_i - \mu_{y_i}) x_j}{\sigma_{y_i}^2} \tag{6.27}$$

where η is the learning rate. Ma and Likharev [155] show that this follows the gradient of r, the reward given to the neuron. Often a simplified form is used by incorporating $\sigma_{y_i}^2$ into the learning rate:

$$\Delta \mu_{ij} = \eta r (y_i - \mu_{y_i}) x_j. \tag{6.28}$$

[155] also investigate an additional term which they say enables the rule to escape from local optima:

$$\Delta \mu_{ij} = \eta [r(y_i - \mu_{y_i}) + \lambda(1 - r)(-y_i - \mu_{y_i})] x_j \tag{6.29}$$

[155] denotes (6.28) and (6.29) by rules A1 and A2 respectively.

In the first set of learning rules, the neuron's output is a function of two independent sets of variables, the inputs and the weights. In the second view, the input is deemed to be fixed and only the synapses are thought of as stochastic. Then $p(w_{ij}) = Z \exp\left[-\frac{(w_{ij} - \mu_{ij})^2}{2\sigma_{ij}^2}\right]$ which leads to a second set of learning rules:

$$\Delta \mu_{ij} = \eta r (w_{ij} - \mu_{ij}) \tag{6.30}$$

which again may include the local optima avoidance term to give

$$\Delta \mu_{ij} = \eta [r(w_{ij} - \mu_{ij}) + \lambda(1 - r)(-w_{ij} - \mu_{ij})] \tag{6.31}$$

[155] denotes (6.30) and (6.31) by rules B1 and B2 respectively. These rules have been tested on a number of standard supervised-learning problems.

6.2.3 Temporal Difference Learning

The method of temporal differences (TD) learning, as was first formalized by Sutton [209], is part of a bigger class of techniques for reinforcement learning. Sutton and Barto [211] have pointed out that reinforcement learning is a learning method which maximizes future reward based on sequences of state-action pairs and rewards from the environment. In [211, 30], reinforcement learning has been thought of as being a close relative of Markov decision processes in that, in reinforcement learning, the agent interacting with the environment can be modeled as a Markov Decision Process (MDP) [24], which consists of

- An initial state s_0
- A state space, S and a action space A
- The probability $P(s_{t+1} = j | s_t = i, a_t = a)$, denoted by $\mathbf{P}_{ij}(a)$, that action a in state i at time t leads to state s_j at time $t+1$
- The probability $P(r_t|s_t, a_t), s_t \in S, r_t \in \Re, a_t \in A$ of getting reward r_t.

A policy π is a mapping from the states to actions, and the goal of a MDP is to find a policy π which would maximize the total expected reward received during the task processing, which is, typically, the discounted sum of rewards under a discounting factor γ:

$$E\{R_t\}_\pi = E\{r_t + \gamma r_{t+1} + \gamma^2 r_{t+2} + \ldots\}_\pi = E\{\sum_{k=0}^{\infty} \gamma^k r_{t+k}\} \tag{6.32}$$

where $\gamma \in [0, 1]$ and the discounting is required so that the sum of an infinite number of discounted rewards is finite and $E\{R_t\}_\pi$ is called *expected discounted return.* After taking the current action a in the current state s and following the policy π, the expected discounted return can be calculated thereafter:

$$Q^\pi(s, a) = E\{\sum_{k=0}^{\infty} \gamma^k r_{t+k} | s_t = s, a_t = a\}_\pi \tag{6.33}$$

$$= E\{r_t + \gamma \sum_{k=0}^{\infty} \gamma^k r_{t+k+1} | s_t = s, a_t = a\}_\pi \tag{6.34}$$

$$= \sum_j \mathbf{P}_{sj}(a)[r_s(a) + \sum_b P(j, b)\gamma Q^\pi(j, b)] \tag{6.35}$$

where Q^π is called the *action-value function.* $\mathbf{P}_{sj}(a)$ is the probability of changing the state from s to j by taking action a, and $P(j, b)$ is the probability of taking action b when in state j. Given an arbitrary initial Q_0^π, the iterative update can be defined by

$$Q_{k+1}^\pi(s, a) = \sum_j \mathbf{P}_{sj}(a)[r_s(a) + \sum_b P(j, b)\gamma Q_k^\pi(j, b)] \tag{6.36}$$

which is called *iterative policy evaluation.*

Lovejoy [153] has solved fully and partially observable Markov decision processes by dynamic programming methods. Sutton [211] has pointed out TD learning is a combination of Monte Carlo ideas [164, 80] and dynamic programming (DP) [23] ideas. TD learning resembles a Monte Carlo method because it learns by sampling the environment according to some policy. TD learning is related to dynamic programming techniques, since its update estimates are based in part on other learned estimates, without waiting for a final outcome. As described in [211], at each time $t+1$, an immediate target for the TD update is formed as $r_{t+1}+\gamma V(s_{t+1})$ and the state value is updated based on the observed reward r_{t+1} and the estimate $V(s_{t+1})$, which is the simplest TD method, known as TD(0). The update rule is given by

$$V(s_t) \leftarrow V(s_t) + \alpha[r_{t+1} + \gamma V(s_{t+1}) - V(s_t)] \tag{6.37}$$

Different from the target for the Monte Carlo update which is the actual return, the target for the TD update is $r_{t+1}+\gamma V(s_{t+1})$. There are two main forms of temporal difference learning for policy control:

- Sarsa learning [210] computes the expected return based on the state-action pairs which gives the state-action value function for the current policy π. The algorithm uses every element of the quintuple of events, $(s_t, a_t, r_{t+1}, s_{t+1}, a_{t+1})$, which describes taking the action a_t in the state s_t at time t, which leads to the results of reward r_{t+1}, and then moving from state s_t to state s_{t+1} at time $t+1$ where an action a_{t+1} will be chosen. Since the relationship between states and state-action pairs is symmetrical, the sarsa algorithm involves transitions from state-action pair to state-action pair, and learns the value of state-action pairs using

$$Q(s_t, a_t) \leftarrow Q(s_t, a_t) + \alpha[r_{t+1} + \gamma Q(s_{t+1}, a_{t+1}) - Q(s_t, a_t)] \tag{6.38}$$

  The general form of the Sarsa algorithm is to first arbitrarily initialize $Q(s,a)$. In each iteration, choose a in the current s using policy π derived from Q, then observe the reward r and the new state $s^{'}$. Then choose $a^{'}$ from the new state $s^{'}$ using policy derived from Q, after which calculate the state-action value by (6.38). Lastly, set $s \leftarrow s^{'}, a \leftarrow a^{'}$.

  Singh [205] has proved that such algorithms converge to the optimal value function under the current policy, assuming that all the states can be visited an infinite number of times.
- Q-learning [225, 226] directly approximates Q^*, the optimal action-value function, by the learned action-value function, Q, by choosing the best possible action in the subsequent state:

$$Q(s_t, a_t) \leftarrow Q(s_t, a_t) + \alpha[r_{t+1} + \gamma \max_a Q(s_{t+1}, a) - Q(s_t, a_t)] \tag{6.39}$$

  in which the policy has the effect that it determines which state-action pairs are visited and updated.

The general form of the Q-learning algorithm is to first arbitrarily initialize $Q(s, a)$. In each iteration, choose the appropriate a in the current s using policy π derived from Q, then observe the reward r and the new state s' and calculate the state-action value by (6.39). Lastly, set $s \leftarrow s'$.

However, if the number of the states and actions in discrete state and action spaces is too large to represent the action value in a state-action table, an alternative way is to use function approximation that is differentiable with respect to some parameters. Considering an approximate value function $V(s)$ parameterized by parameters θ whose the inputs are states and the target function is the true value function, V^*, the function approximation method is performed to minimize the mean-squared error (MSE) over some distribution, P, between the true value function, $V^*(s)$ and the approximate value function $V(s)$:

$$MSE(\theta) = \sum_{s \in S} \mathbf{P}(s)[V^*(s) - V(s)]^2 \tag{6.40}$$

$\mathbf{P}(s)$ is the stationary probability of state s after any burn-in or transient period. With the same distribution, P, for gradient-descent TD(λ) in [211], the update rule for the parameters θ is given by

$$\Delta\theta = \alpha\delta_t e_t, \tag{6.41}$$

where δ_t is called *TD error* and defined as

$$\delta_t = r_{t+1} + \gamma V_t(s_{t+1}) - V_t(s_t), \tag{6.42}$$

and e_t is a column vector of eligibility traces defined as,

$$e_t = \gamma\lambda e_{t-1} + \nabla_{\theta_t} V_t(s_t) \tag{6.43}$$

Function approximation has been widely studied in reinforcement learning. Persson [5] has used temporal difference learning techniques in combination with artifical neural networks to solve strategic control problems. Ackley and Littman [2] derived a complementary reinforcement backpropagation algorithm that consists of a feed-forward network mapping an encoding of the state to an encoding of the action. Sallans and Hinton [196] use a product of experts model to approximate the values of states and actions in a Markov decision process.

6.2.4 Evolutionary Algorithms for Reinforcement Learning

While reinforcement learning with temporal difference learning attempts to learn the value function $V^{\pi*}$ through experience and there is not an explicit representation of a policy, reinforcement learning with evolutionary algorithm is a policy-space search that maintains explicit representations of policies and modifies them through evolutionary algorithms.

Evolutionary Algorithms (EAs) [116, 104, 139] are global search techniques derived from Darwin's theory of evolution by natural selection. An EA iteratively updates a population of encoded potential solutions that are called *chromosomes*. During each iteration, also called a *generation*, the EA evaluates solutions based on the fitness of each solution and generates offspring accordingly. Thus, good fitness improves the probability of being selected and only those selected solutions are then modified through genetic process such as mutation and combined to give better solutions in subsequent iterations.

```
procedure EARL-1
begin
    t = 0;
    initialize a population of policies, P(t);
    evaluate policies in P(t);
    while termination condition not satisfied do
        begin
            t = t + 1;
            select high-payoff policies, P(t), from policies in P(t-1);
            update policies in P(t);
            evaluate policies in P(t);
        end
end.
```

Fig. 6.7. Pseudo-code for Evolutionary Algorithm Reinforcement Learning system [172]

Moriarty et al. [172] gave a concrete example of a simple evolutionary reinforcement learning algorithm as shown in Figure 6.7, where a possible policy is represented by a single chromosome with a single gene associated with each observed state. Each gene's value corresponds to the action value associated with the state. There are no constraints on how the fitness of an individual policy is evaluated, however, the fitness of each chromosome would reflect the expected accumulated fitness for an agent with the given policy. In accordance with the determined fitness of all policies in the population, a new population of policies is generated with the usual evolutionary algorithm where a selection method is first performed to probabilistically select individuals based on relative fitness. The selected parent policies are cloned to form the offspring and the expected number of offspring for a policy is usually proportional to its fitness. Then a new population of policies is generated by applying the standard genetic operation of crossover and mutation to the clones. Thus the process of generating new populations can continue for a fixed number of iterations until reaching an acceptable level of performance.

[172] has pointed out that one of the most critical features that distinguishes EA-based reinforcement learning from TD-based reinforcement learning is the policy explicitly defined. EA-based reinforcement learning is a policy-space search method that modifies the explicitly defined policies through a variety of search operators, however, TD-based reinforcement learning is a value function method that attempts learning the value function which returns the expected long-term reward for the optimal policy without a policy explicitly defined. Generally speaking, there are ways to represent the policies [172], such as the rule-based policies or neural network representation of policies. In the first, Grefenstette et al. [107] have represented the policy as a set of condition-action rules where an action matches a set of states. Policies can also be represented by neural networks [228, 236, 208, 246]. An agent's decision policy is expressed by a sequence of real-valued weights, where EAs can be applied to optimize the weights of the neural network.

6.3 Clustering with Reinforcement Learning

In this section, we investigate using reinforcement learning with reward functions which are related to our previous clustering methods in that the reward functions also allow the reinforcement learning algorithms to overcome the disadvantages of a poor initialization and achieve the globally optimal clustering.

We applied the Bernoulli algorithm [149] to the artificial data set shown in Figure 6.8, left, but the Bernoulli algorithm failed to identify all the clusters successfully as shown in Figure 6.8, right.

The Bernoulli algorithm also suffers from a poor initialization and shows sensitivity to the prototypes' initialization. This results in dead prototypes and convergence to a local optimum. The main reason for these problems is that we update in this algorithm the winner prototypes only, not all of them. Poor initialization may allow some prototypes to learn while others don't respond or learn at all. Thus we illustrate three different reward functions [15] that overcome the problem of poor initialization and the convergence to

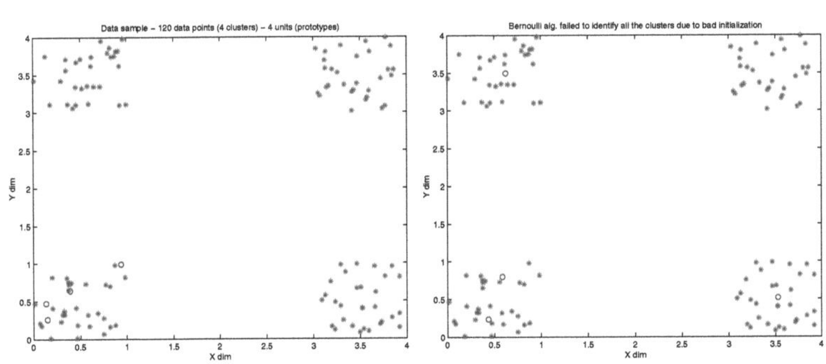

Fig. 6.8. Left: artificial data set is shown as 4 clusters of red '*'s, and 4 prototypes of blue 'o's. Right: Bernoulli algorithm failed to identify all the clusters successfully.

the local optimum. These new reward functions allow all the nodes to learn, not only the winners.

6.3.1 New Algorithm RL1

The RL1 algorithm has the following reward function which is motivated by inverse weighted clustering algorithm (IWC):

$$r_i = \begin{cases} \frac{\|\mathbf{x}-\mathbf{m}_{k*}\|^3}{\|\mathbf{x}-\mathbf{m}_i\|^3} & \text{if } y_i = 1 \\ -\frac{\|\mathbf{x}-\mathbf{m}_{k*}\|^3}{\|\mathbf{x}-\mathbf{m}_i\|^3} & \text{if } y_i = 0 \end{cases} \tag{6.44}$$

where

$$k* = \arg\min_{k=1}^{K}(\| \mathbf{x} - \mathbf{m}_k \|) \tag{6.45}$$

This new reward function has the following advantages:

1. We are rewarding all prototypes (nodes), not only the winners and thus all prototypes will learn to find the clusters even if they are initialized badly.
2. This reward function allows the prototypes to respond differently to each other, and each prototype, before moving to any new locations, responds to all the other prototypes' positions, and hence it is possible for it to identify the free clusters that are not recognized by the other prototypes.
3. This reward function gives the highest value, 1, for the highest similarity between the data point and the node (prototype).

We have better results by changing the power degree in (6.44) from 2 to a small value e.g. 0.1, which will extend the influence of distances between the units and data points.

Simulations

Figure 6.9 shows the result after applying the RL1 algorithm to the artificial data set, but with very poor prototypes' initialization.

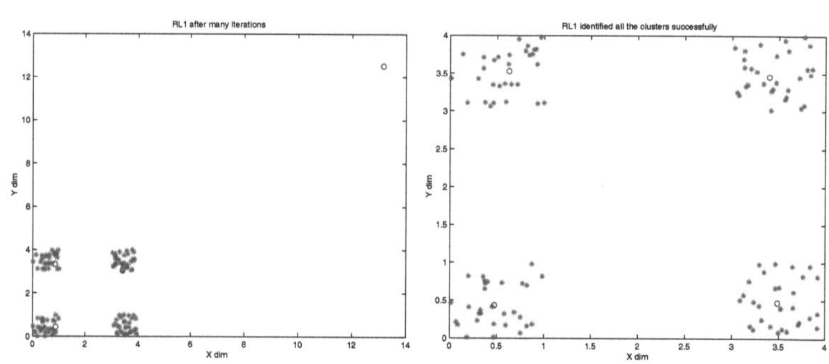

Fig. 6.9. Left: RL1 result after many iterations but before convergence. Right: RL1 result after convergence.

Figure 6.9, left, shows the prototypes after many iterations but before convergence; in this Figure, we can see one prototype still distant from the data points while others have spread into the data; this distant prototype still has the ability to learn even if it is very far from data, and this is an advantage for this algorithm over the previous algorithms.

6.3.2 New Algorithm RL2

The RL2 algorithm has the following reward function which is motivated by inverse exponential K-means algorithm 1 (IEK1):

$$r_i = \begin{cases} 1 & \text{if } i = k^* \text{ and } y_i = 1 \\ \frac{1-\exp(-\beta\|\mathbf{x}-\mathbf{m}_{k*}\|^3)}{\|\mathbf{x}-\mathbf{m}_i\|^3} & \text{if } i \neq k^* \text{ and } y_i = 1 \\ -1 & \text{if } i = k^* \text{ and } y_i = 0 \\ \frac{\exp(-\beta\|\mathbf{x}-\mathbf{m}_{k*}\|^3)-1}{\|\mathbf{x}-\mathbf{m}_i\|^3} & \text{if } i \neq k^* \text{ and } y_i = 0 \end{cases} \tag{6.46}$$

where again $k* = \arg\min_j \| \mathbf{x} - \mathbf{m}_j \|$.

The reward function (6.46) has values ranged between 0 and 1. We need to update the closest prototype (or most similar one) by giving it directly the maximum possible reward value, which equals 1, to allow it to learn more than others and also to avoid any division by zero, which may happen using the second equation in (6.46). The second equation in (6.46) is used for all the other prototypes. Prototypes closer (or more similar) to the input data sample will learn more than others by taking a higher reward value, thus all prototypes learn appropriately.

6.3.3 New Algorithm RL3

The RL3 algorithm has the following reward function which is motivated by K-Harmonic means:

$$r_i = \begin{cases} \frac{1}{\|\mathbf{x}-\mathbf{m}_i\|^4\{\sum_{l=1}^{K}\frac{1}{\|\mathbf{x}-\mathbf{m}_l\|^2}\}^2} & \text{if } y_i = 1 \\ \frac{-1}{\|\mathbf{x}-\mathbf{m}_i\|^4\{\sum_{l=1}^{K}\frac{1}{\|\mathbf{x}-\mathbf{m}_l\|^2}\}^2} & \text{if } y_i = 0 \end{cases} \tag{6.47}$$

The reward function in (6.47) has similar principles to the previous new reward functions. It has values ranged between 0 and 1. All the prototypes can learn in an effective way. The prototype that is more similar to the input data sample gets a higher reward value.

In implementation, to avoid any division by zero we can rewrite (6.47) as follows:

$$r_i = \begin{cases} \dfrac{\frac{\|\mathbf{x}-\mathbf{m}_{k*}\|^4}{\|\mathbf{x}-\mathbf{m}_i\|^4}}{\left\{1+\sum_{l\neq k*}^{K}\frac{\|\mathbf{x}-\mathbf{m}_{k*}\|^2}{\|\mathbf{x}-\mathbf{m}_l\|^2}\right\}^2} & \text{if } y_i = 1 \\ \\ \dfrac{-\frac{\|\mathbf{x}-\mathbf{m}_{k*}\|^4}{\|\mathbf{x}-\mathbf{m}_i\|^4}}{\left\{1+\sum_{l\neq k*}^{K}\frac{\|\mathbf{x}-\mathbf{m}_{k*}\|^2}{\|\mathbf{x}-\mathbf{m}_l\|^2}\right\}^2} & \text{if } y_i = 0 \end{cases} \tag{6.48}$$

where $k* = \arg\min_j \parallel \mathbf{x} - \mathbf{m}_j \parallel$.
Notice: $\frac{\|\mathbf{x}-\mathbf{m}_{k*}\|^4}{\|\mathbf{x}-\mathbf{m}_{k*}\|^4}$ is always set to 1.

6.3.4 Simulations

Figure 6.10 shows the results after applying the Bernoulli algorithm, top right, RL2, bottom left, and RL3, bottom right, to the artificial data set shown in Figure 6.10, top left. RL2 and RL3 succeed in identifying the clusters successfully while the Bernoulli model failed.

To show how these algorithms behave with dead prototypes, we have Figure 6.11, left, which contains some. Figure 6.11, right, shows the result after

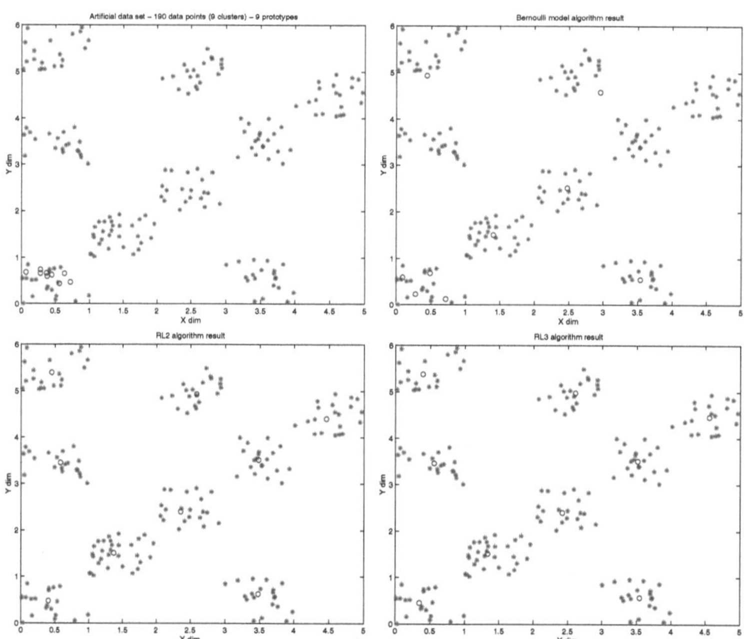

Fig. 6.10. Top left: artificial data set with poor prototypes' initialization. Top right: the Bernoulli algorithm result. Bottom left: RL2 algorithm result. Bottom right: RL3 algorithm result.

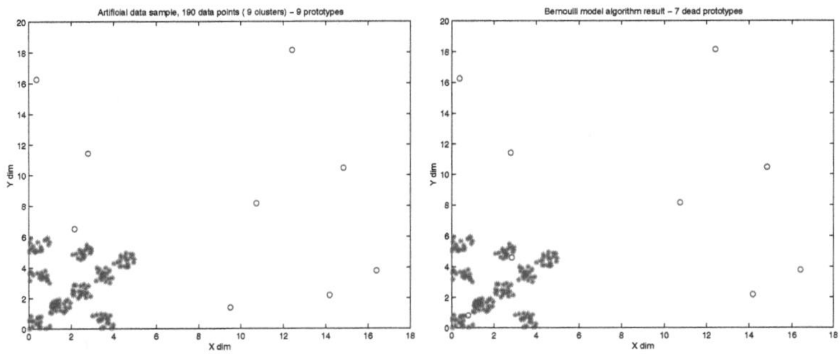

Fig. 6.11. Left: artificial data set with poor prototypes' initialization. Right: Bernoulli algorithm result, failed to identify clusters - 7 dead prototypes.

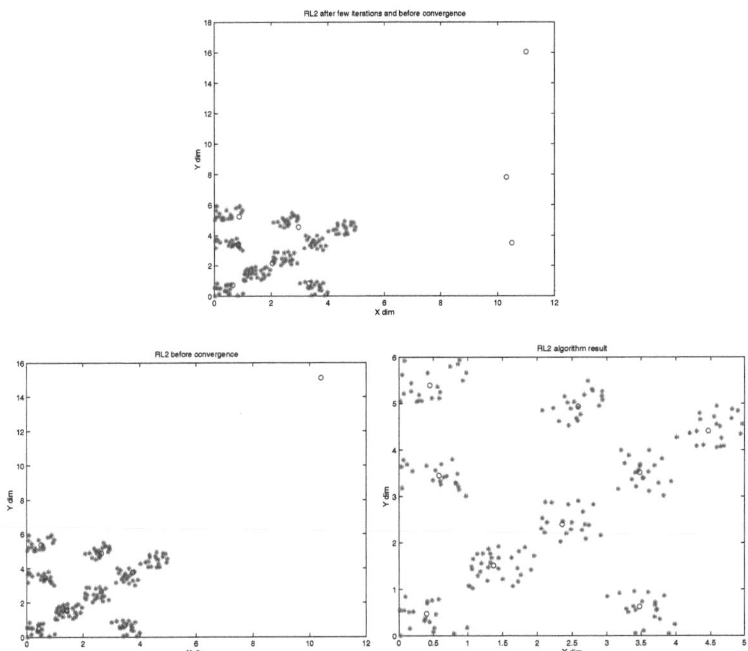

Fig. 6.12. RL2 algorithm results. Top: result after 10 iterations and before convergence. Bottom left: result after 30 iterations and before convergence. Bottom right: RL2 algorithm result after 70 iterations.

applying the Bernoulli algorithm to the same artificial data set as Figure 6.10, top left, but with very poor prototypes' initialization as shown in Figure 6.11, left. The Bernoulli algorithm gave bad results and there are 7 dead prototypes which don't learn. Figure 6.12 shows the result after applying the RL2 algorithm to the same artificial data set. From Figure 6.12, top and bottom left, we can see some prototypes still distant from the data points while others spread into data. These distant prototypes still have the ability to learn and identify clusters, as shown in Figure 6.12, bottom right, even if they are far from data. This activation of dead prototypes is an advantage of these new algorithms.

6.3.5 Topology Preserving Mapping

Again we use the same structure as the GTM to visualize data, but this time with RL1 and RL2 for the learning process to construct RL1ToM and RL2ToM.

Artificial data set

We create a simulation with 20 latent points deemed to be equally spaced in a one dimensional latent space, passed through 5 Gaussian basis functions and then mapped to the data space by the linear mapping W which is the only parameter we adjust. We generated 500 two dimensional data points, (x_1, x_2), from the function $x_2 = x_1 + 1.25\sin(x_1) + \mu$ where μ is noise from a uniform distribution in [0,1]. Final result from the RL1ToM is shown in Figure 6.13.

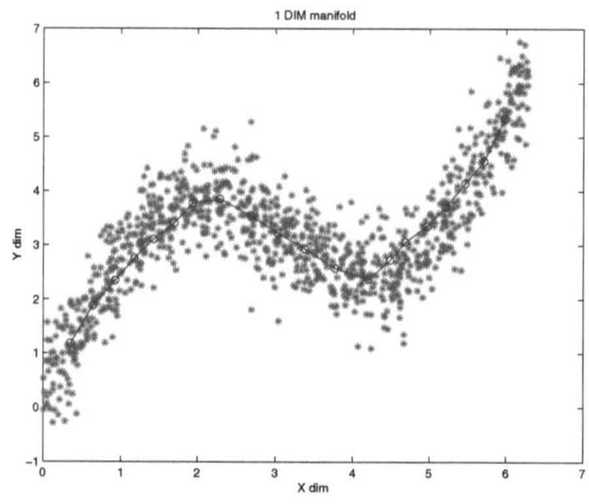

Fig. 6.13. The resulting prototypes' positions after applying RL1ToM. Prototypes are shown as blue 'o's.

Real data sets

We show in Figure 6.14 and Figure 6.15 the projections of the real data sets, iris, algae, genes and glass, onto a two dimensional grid of latent points (10 x 10) using RL1ToM and RL2ToM, respectively. The final results are comparable with others we have with these data sets from a variety of different

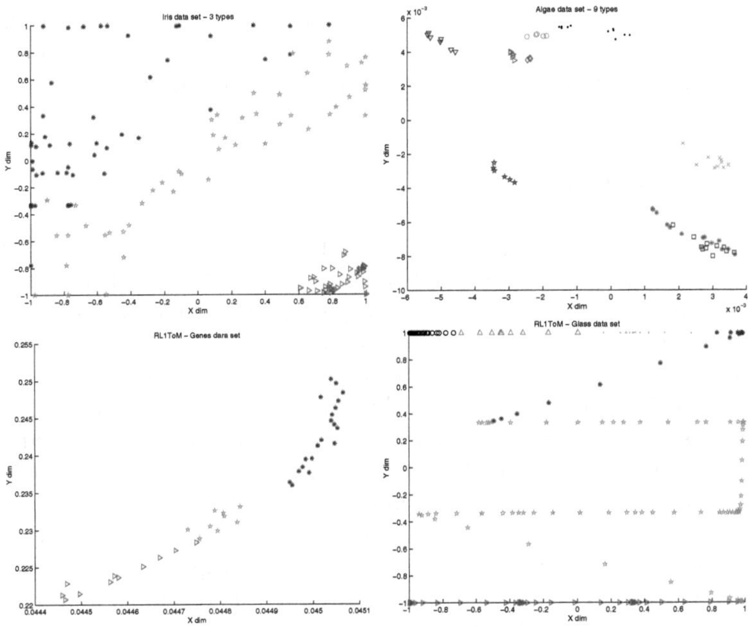

Fig. 6.14. Visualisation using the RL1ToM on 4 real data sets

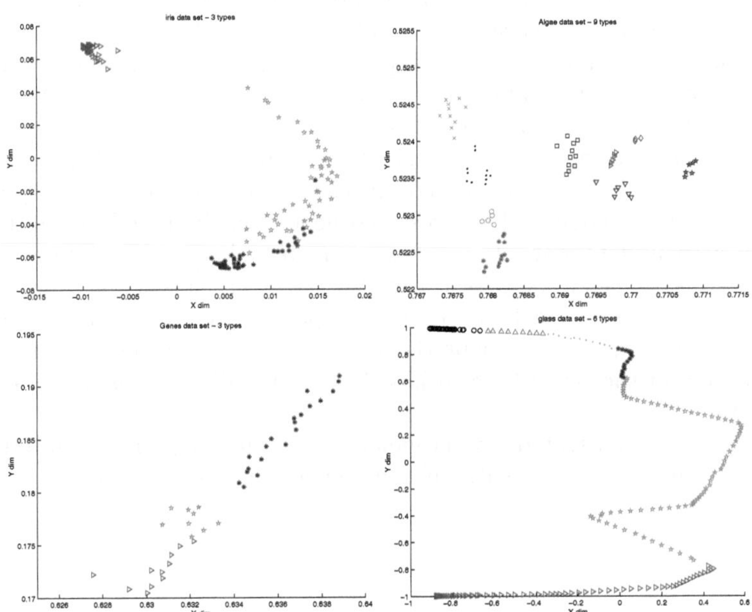

Fig. 6.15. Visualisation using the RL2ToM on 4 real data sets

algorithms. The RL2ToM algorithm gives the best result, similar to those of IKoToM, for the algae data set. It succeeds in visualizing all the 9 clusters successfully, while most of the previous methods failed to identify all of them.

6.4 Conclusion

We have discussed one shortcoming of the K-means algorithm: its sensitivity to poor initialization which leads it to converge to a local rather than global optimum. We have shown how different performance functions lead to algorithms which incorporate the correct mixture of local and global knowledge to allow prototypes to optimally cluster a data set. We have derived both a batch algorithm and online algorithms from these performance functions. We emphasise that these performance functions are the same as used previously however instead of creating a batch algorithm we now have online or incremental updating of the parameters.

We have extended these algorithms by using them with an underlying latent space which enables topology preserving mappings to be developed. We have illustrated these mappings (both batch and online versions) on a variety of data sets and shown how they may be used to visualise these data sets.

We have shown how reinforcement learning of cluster prototypes can be performed robustly by altering the reward function associated with finding the clusters. We have illustrated three different reward functions which clearly have a family resemblance. Most importantly all three overcome the disadvantages of poor initialization in that they do not succumb to local minima as the existing Bernoulli algorithm does.

It is worth stating that the original Bernoulli algorithm [149] seems to somewhat cheat in its formulation of the clustering method: we already know that the winning node is the closest and hence that it should be emitting 1 and so every update moves the node's parameters closer to the data point for which it was deemed to be closest. However the algorithms herein sample each and every node using the distribution corresponding to its current parameters i.e. no prior assumptions about the winning node are made except in the formulation of the reward which has the winning distance as numerator of the reward.

We have also illustrated how a topology preserving mapping can be created by using these algorithms with an underlying fixed latent space.

7

Connectivity Graphs and Clustering with Similarity Functions

7.1 Different Similarity Graphs (or Connectivity Graphs)

To construct a similarity graph we transform a given set $\mathbf{x}_1, ..., \mathbf{x}_n$ of data points with pairwise similarities s_{ij} or distances d_{ij} into a graph. There are several popular methods to construct similarity graphs [154]. The goal of constructing similarity graphs is to model the local neighborhood relationships between data points. In this section we review two popular methods to construct a similarity graph, and then we introduce a new algorithm that solves some problems that can not be solved by the others.

7.1.1 The ε-Neighborhood Graph

In this algorithm we connect all points whose pairwise distances are smaller than ε. Choosing a useful value of ε is not a trivial task [154]. Sometimes it is not possible to find a suitable value of ε if we have data on different scales or densities.

7.1.2 k-Nearest Neighbor Graphs

In this algorithm we connect each point to all points among its k-nearest neighbors. This leads to a directed graph as the neighborhood relationship is not necessarily symmetric. To make this graph undirected we have two choices. The first one is to simply ignore the direction of the edges, so that points $\mathbf{x}$ and $\mathbf{y}$ are connected with undirected edge if $\mathbf{x}$ is among the k-nearest neighbors of $\mathbf{y}$, or if $\mathbf{y}$ is among the k-nearest neighbors of $\mathbf{x}$. The resulting graph is usually called k-nearest neighbor graph. The second way is to connect points $\mathbf{x}$ and $\mathbf{y}$ if both of them are among the k-nearest neighbors of the other. The resulting graph is called mutual k-nearest neighbor graph [154].

W.A. Barbakh et al.: Non-Stand. Param. Adapt. for Explor. Data Analys., SCI 249, pp. 109–122.
springerlink.com

7.1.3 New Similarity Graph

We show here a new similarity graph that can be used for clustering and spectral clustering. In this algorithm [12, 18], we connect all points whose pairwise weights are higher than a useful threshold. This threshold is a value ranging between 0 and 1; if its value equals 0, all the points will be connected together. If its value equals 1, only the points which are close to each other will be connected. The weight between vertex $\mathbf{v}_i$ and $\mathbf{v}_j$ is calculated by

$$weight(\mathbf{v}_i, \mathbf{v}_j) = \frac{min(X, Y)}{\parallel \mathbf{v}_i - \mathbf{v}_j \parallel} \tag{7.1}$$

where X is the distance between $\mathbf{v}_i$ and the closest vertex to it.

Y is the distance between $\mathbf{v}_j$ and the closest vertex to it.

(7.1) has the same motivation as our clustering algorithms in [13, 16, 17], see (4.26) which is used to measure the similarity between a data point and a prototype. In (7.1), we use the same idea to measure the similarity between two data points.

This new similarity graph algorithm solves some problems that can not be solved by the previous algorithms as we will see in the next example. However this new algorithm is not designed to replace the previous algorithms but to add new behavior to them and to give better results for different applications.

Example 1:
Suppose that we have an artificial data set like that shown in Figure 7.1. $\mathbf{x}_1$ and $\mathbf{x}_2$ represent the first diffuse cluster, and $\mathbf{x}_3$ and $\mathbf{x}_4$ represent the second tighter cluster. Figure 7.2 shows the results after applying ε-neighborhood graph, with different values of ε (ε=0.5, 1.1, 1.5 and 2.1 respectively), to the artificial data set shown in Figure 7.1. From Figure 7.2 we can see it is not

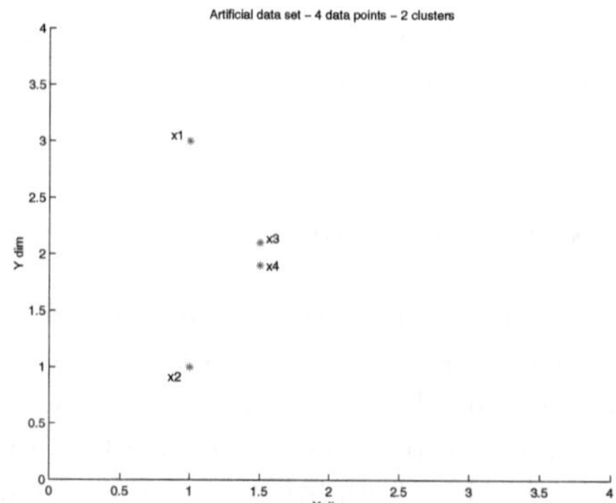

Fig. 7.1. Artificial data set is shown as 2 clusters of red '*'s

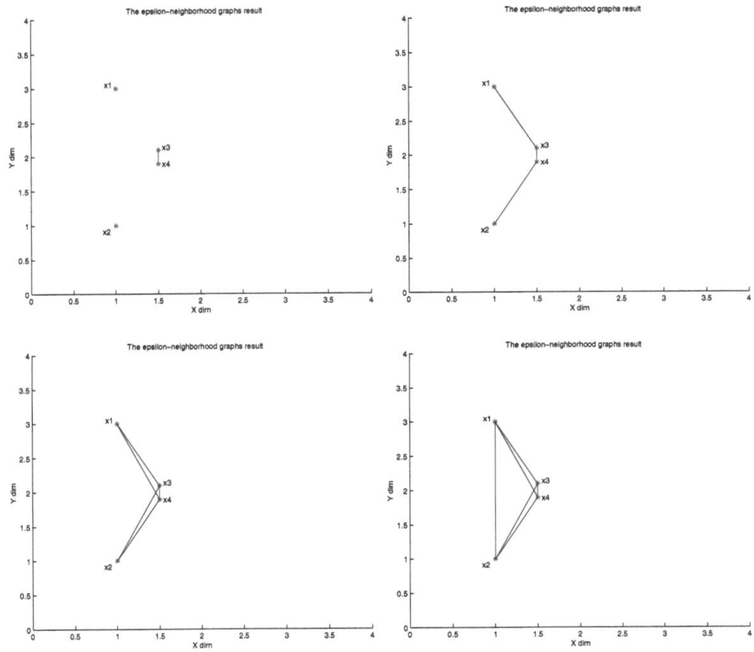

Fig. 7.2. The ε-neighborhood graph results with different values of ε (ε=0.5, 1.1, 1.5 and 2.1 respectively)

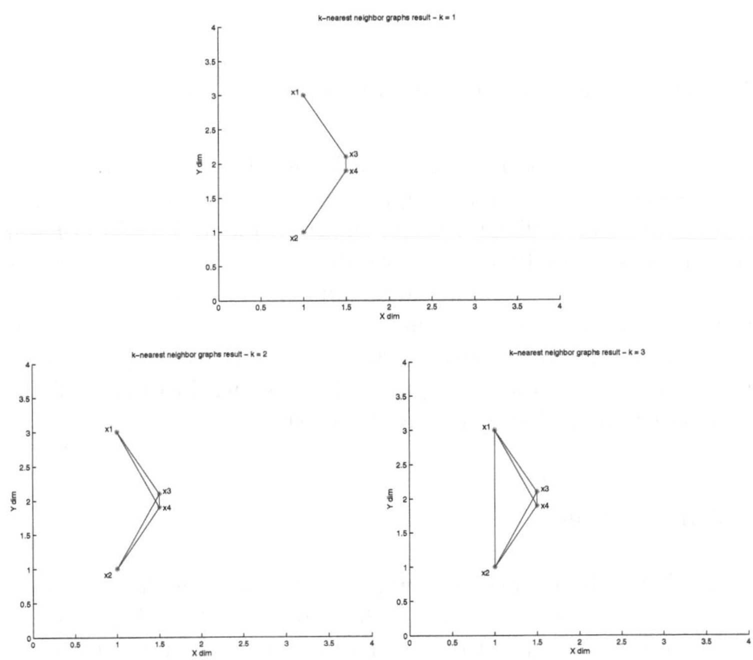

Fig. 7.3. The k-nearest neighbor graphs results with k=1, 2 and 3

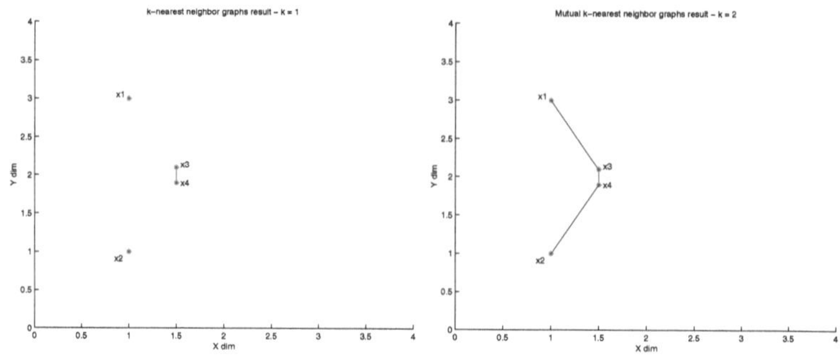

Fig. 7.4. The mutual k-nearest neighbor graphs results with k=1 and 2

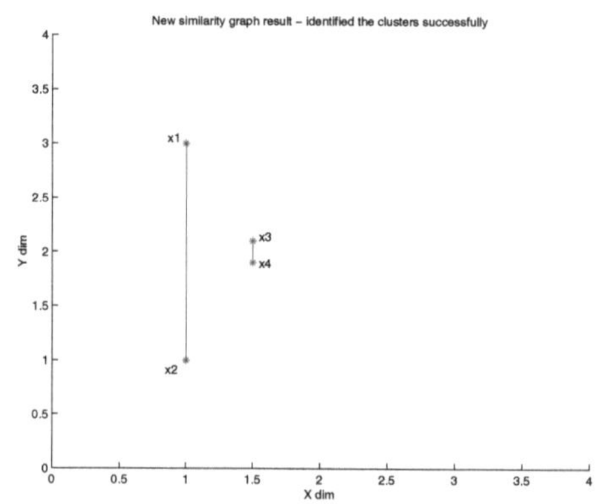

Fig. 7.5. The new similarity graph algorithm succeeds in identifying the clusters

possible to find a useful parameter ε that defines the two clusters successfully. Figure 7.3 shows the results after applying k-nearest neighbor graphs, with k=1, 2 and 3, to the artificial data set shown in Figure 7.1. From Figure 7.3 we can see there is no useful parameter k that defines the two clusters successfully. Figure 7.4 shows the results after applying mutual k-nearest neighbor graphs, with k=1 and 2, to the artificial data set shown in Figure 7.1. Figure 7.5 shows the result after applying the new similarity graph algorithm to the artificial data set shown in Figure 7.1. As shown in Figure 7.5 the new algorithm succeeds in identifying the two clusters.

7.2 Simulations

Figure 7.7 shows the results after applying the new similarity graph algorithm, with two different threshold values, to the artificial data set shown in Figure 7.6. In Figure 7.7, top, the new algorithm identifies each circle as a cluster. We can get another result by choosing another threshold value as

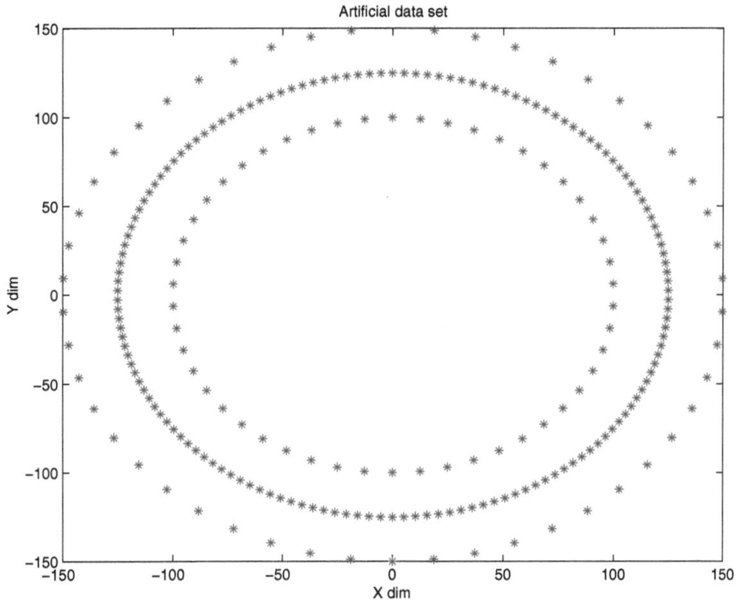

Fig. 7.6. Artificial data set consisting of 3 circles of red '*'s. The outer and inner circles have roughly similar density.

shown in Figure 7.7, bottom. In Figure 7.7, bottom, the outer and inner circles have roughly similar density and are identified as one cluster while the middle circle, which has higher density, is identified as another cluster. This result can not be achieved by either ε-neighborhood graph or k-nearest neighbor graphs.

In Figure 7.8, we have an artificial data set taken from two different Gaussian distributions. Figure 7.9 shows the results after applying the new similarity graph algorithm to the artificial data set shown in Figure 7.8. From Figure 7.9, top, we can see that all the data points from first distribution are connected together without any connection to the points from the second distribution.

7.3 Clustering with Similarity Functions

In this section we show how we can utilise the benefits from the previous algorithms in Chapter 3 and introduce new clustering algorithms that are based on similarity functions. The inverse weighted clustering algorithm (IWC) defined in Chapter 3 has the following rule:

$$\mathbf{m}_k = \frac{\sum_{i=1}^{N} b_{ik}\mathbf{x}_i}{\sum_{i=1}^{N} b_{ik}} \tag{7.2}$$

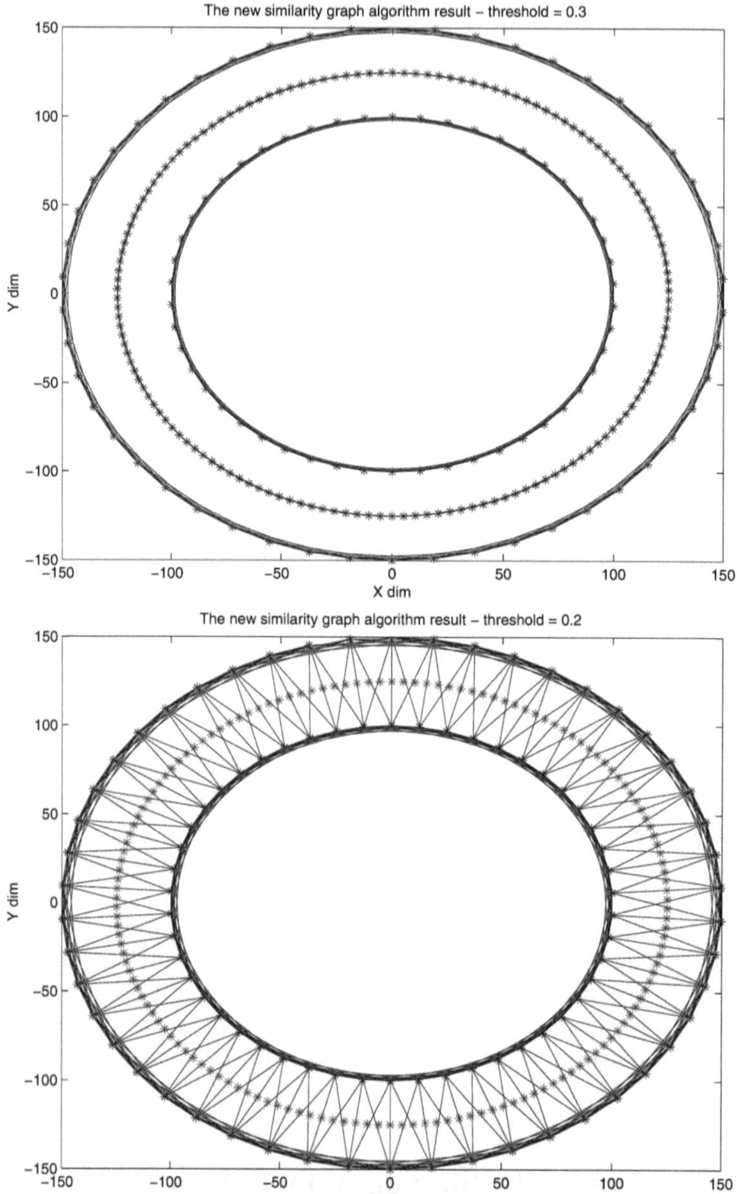

Fig. 7.7. The new similarity graph algorithm results. Top: the new algorithm identifies each circle as a cluster. Bottom: the new algorithm identifies outer and inner circles as one cluster.

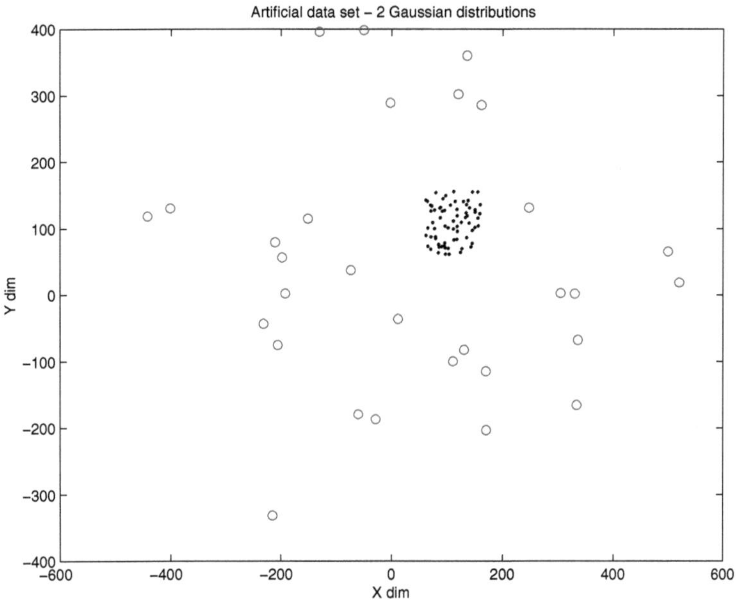

Fig. 7.8. Artificial data set taken from two different Gaussian distributions, the red 'o's represents the first distribution and the black dots represents the second distributions

where

$$b_{ik} = \frac{\| \mathbf{x}_i - \mathbf{m}_{k*} \|^{P+2}}{\| \mathbf{x}_i - \mathbf{m}_k \|^{P+2}} \tag{7.3}$$

(7.2) introduces the clustering algorithm (IWC) that is insensitive to the prototypes' initialization. The question now, is how can we use this algorithm for clustering if we have similarity functions that measure the similarity between data points? How can we work with the similarity function and at the same time get the benefit of (7.2)?

7.3.1 Exponential Function as Similarity Function

We take as an example the exponential function to measure the similarity between data points and prototypes; we can use other similarity functions. We compare the new algorithm with the soft K-means algorithm as it is also based on the exponential function. We show in simulations that the new algorithm gives better results than both the K-means and the soft K-means algorithms. Also, this new algorithm solves the problem of excessive computation in [151]. Let the exponential function be used, as an example, to measure the similarities between points as similarity$(\mathbf{x}, \mathbf{m}) = \exp(- \| \mathbf{x} - \mathbf{m} \|)$.

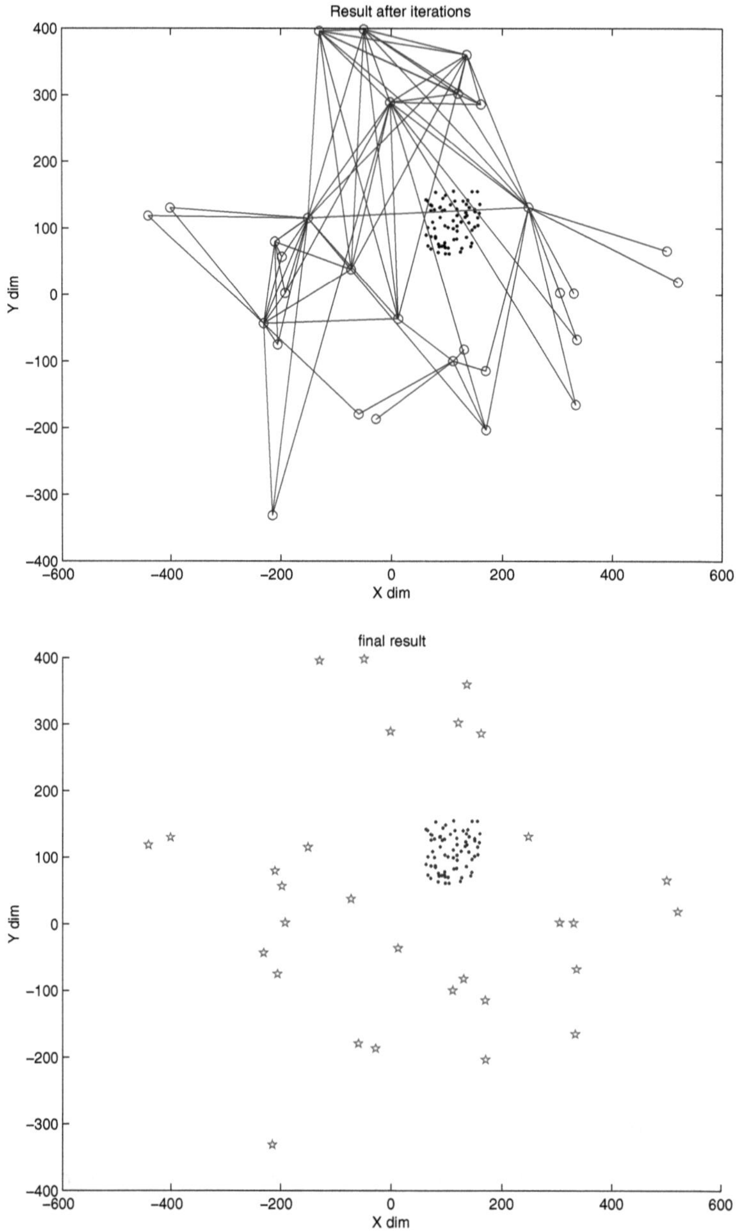

Fig. 7.9. The new similarity graph algorithm results. Top: the data points in the first distribution are connected together. Bottom: final result, clusters identified successfully from the points that are connected together.

To use this similarity function for clustering while taking the benefits of (7.2) we need to go through the following steps:

1. Measure the similarities between all data points and prototypes.
2. For alternative similarity function, if the similarity measurements are outwith [0,1], normalize, so that 1 corresponds to the highest similarity and 0 corresponds to no similarity.
3. Map the similarity measurements to distances:
   Distance $(\mathbf{x}, \mathbf{m}) = 1 - \exp(- \parallel \mathbf{x} - \mathbf{m} \parallel)$.
4. Compute (7.2) for all prototypes using the new distance measurements generated from the similarity function.

$$b_{ik} = \frac{distance(\mathbf{x}_i - \mathbf{m}_{k*})^{P+2}}{distance(\mathbf{x}_i - \mathbf{m}_k)^{P+2}}. \tag{7.4}$$

We will call this Inverse Weighted Clustering with Similarity Function algorithm (IWCwSF) [18].

7.3.2 Simulations

Figure 7.10 shows an artificial data set consisting of 7 clusters of red '*'s, and all the prototypes are initialized to lie within one cluster and are shown as blue 'o's. Figure 7.11 shows the results after applying the soft K-means and

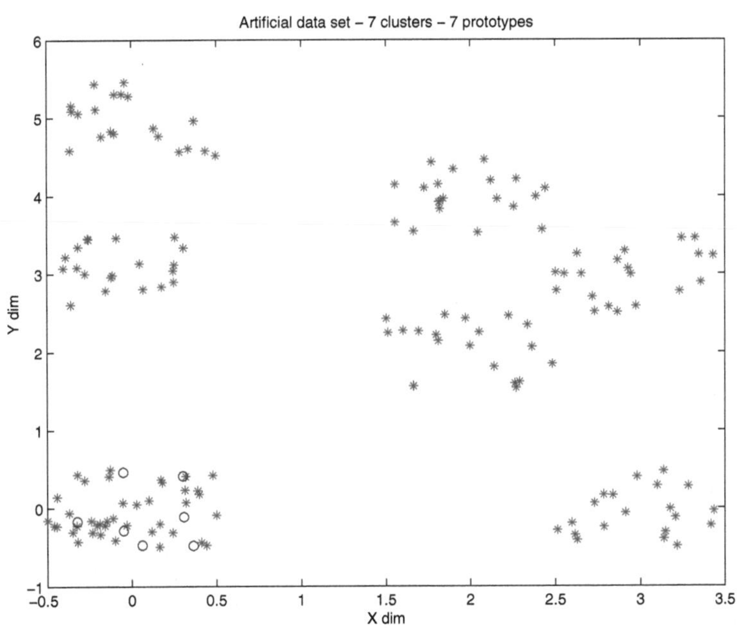

Fig. 7.10. Artificial data set: 7 clusters of red '*'s, and 7 prototypes of blue 'o's

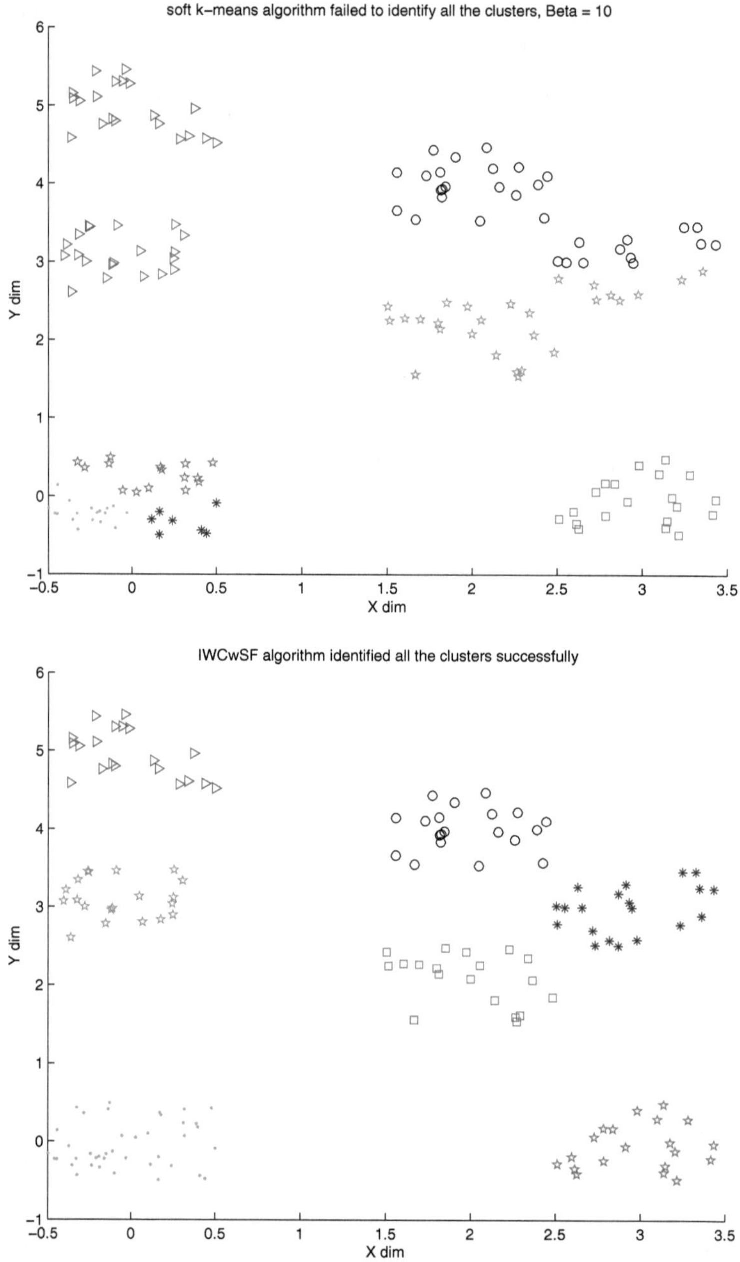

Fig. 7.11. Top: result after applying the soft K-means algorithm. Bottom: result after applying the inverse weighted clustering with similarity function (IWCwSF).

the inverse weighted clustering with similarity function (IWCwSF) to this artificial data set shown in Figure 7.10. In Figure 7.11, top, we see the first cluster, bottom left, is divided into three sub clusters. Also, the upper two clusters, left, are grouped together as one cluster. This poor result from the soft K-means appeared due to its convergence to a local optimum. As shown in Figure 7.11, while the soft K-means failed to identify the clusters, top, the IWCwSF algorithm identified all of them successfully, bottom.

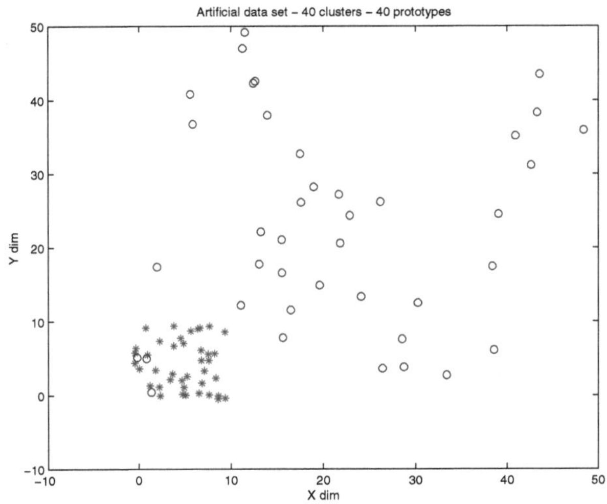

Fig. 7.12. Artificial data set: 40 clusters of red '*'s, and 40 prototypes of blue 'o's

To show how the new algorithm behaves with dead prototypes, we have in Figure 7.12 another artificial data set consisting of 40 data points. Each data point represents a cluster, so we have 40 clusters, and 40 prototypes are initialized randomly and very far from data to represent some dead prototypes. Figure 7.13 shows the result after applying the soft K-means and the IWCwSF to the artificial data set shown in Figure 7.12. Again, the IWCwSF algorithm identified all the clusters successfully, bottom, while the soft K-means algorithm did not, top.

7.3.3 Inverse Weighted Clustering with Similarity Function Topology Preserving Mapping(IWCSFToM)

The IWCSFToM algorithm uses the inverse weighted clustering with similarity function (IWCwSF) algorithm for the learning process. It has the same structure as the GTM, with a number of latent points that are mapped to a feature space by M Gaussian functions, and then into the data space by a matrix W.

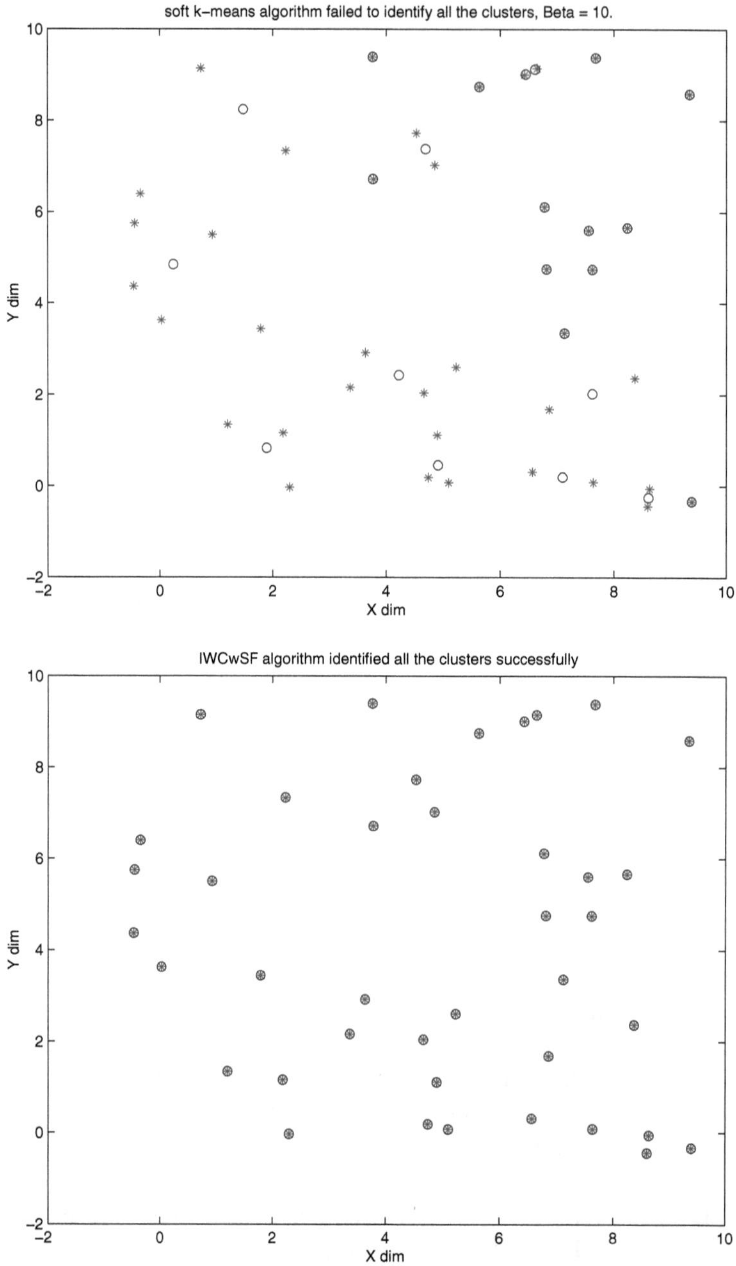

Fig. 7.13. Top: result after applying the soft K-means algorithm to the data of Figure 7.12. Bottom: result after applying the inverse weighted clustering with similarity function (IWCwSF).

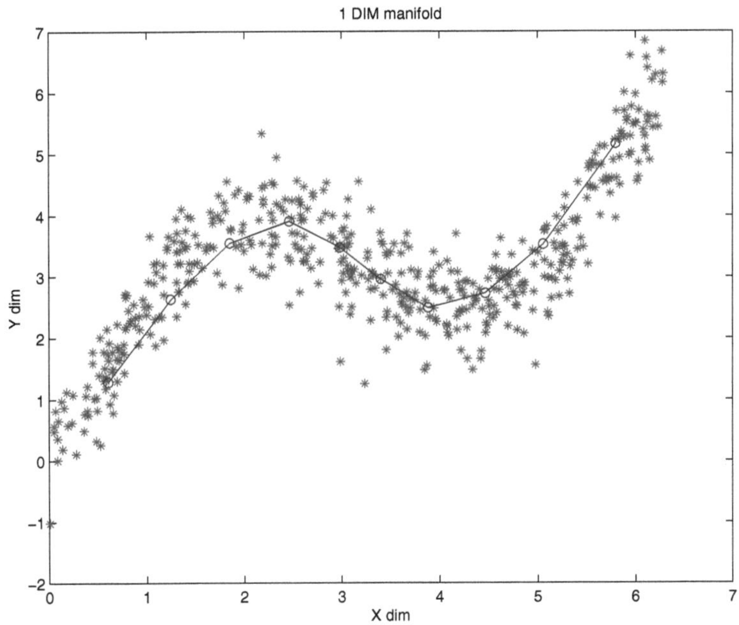

Fig. 7.14. The resulting prototypes' positions after applying IWCSFToM. The prototypes are shown as blue 'o's.

Simulations

1. Artificial data set
   We create a simulation with 10 latent points deemed to be equally spaced in a one dimensional latent space, passed through 5 Gaussian basis functions and then mapped to the data space by the linear mapping W which is the only parameter we adjust. We generated 500 two dimensional data points, (x_1, x_2), from the function $x_2 = x_1 + 1.25\sin(x_1) + \mu$ where μ is noise from a uniform distribution in [0,1]. Final result from the IWCSFToM is shown in Figure 7.14 in which the projections of consecutive latent points are joined. We see that nearby latent points take responsibility for nearby data points.
2. Real data sets
   Figure 7.15 shows the results of applying IWCSFToM to the iris, algae and bank data sets. The real data sets are projected onto a two dimensional grid of latent points (10 x 10). For the iris data set, top , we can see one cluster is distant and separated, while there is difficulty to separate the other two completely. For the algae data set, bottom left, we have 7 clusters separated, and the two clusters to the left are grouped together. For the bank data set, bottom right, the two clusters are identified successfully.

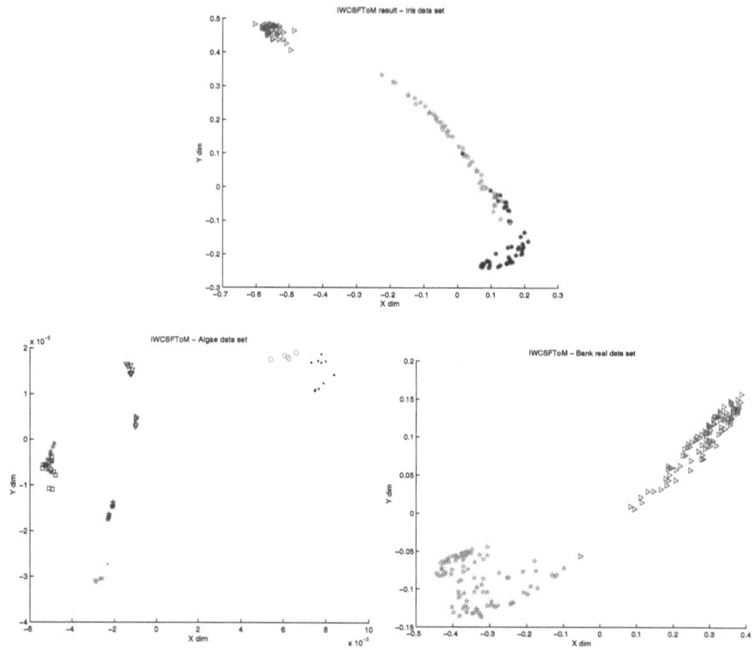

Fig. 7.15. The results of applying the IWCSFToM on the real data sets, iris, top, algae, bottom left and bank, bottom right

7.4 Conclusion

Similarity graphs provide a relatively new way of performing clustering. However we have shown that the standard methods of obtaining clusters from similarity graphs are less than perfect for identification of clusters, particularly when the clusters have different densities.

Thus we have illustrated a new method of identifying clusters using the concepts of similarity graphs. However unlike the standard methods, our method takes into account, not just the distance between pairs of data points, but also the shortest distance between either data point and any other data point. In doing so, we can account for variable density of data points within a data set. Initial results on artificial data sets are very promising. We note that this combination of local and global information sits well with the general proposals within this book which is that this combination provides a robust and general method for reliable clustering.

We have shown how to get the benefits from the new algorithms shown previously to generate new clustering algorithms based on similarity functions: again these functions provide a mixture of local and global information by utilising the closest prototype to the current data as well as the prototype whose position we are adjusting.

Finally we have extended one of these algorithms for visualization by using the same underlying latent space as was used in previous chapters.

8

Reinforcement Learning of Projections

In this chapter, we derive three general reinforcement learning methods [234]. for projection problems under the framework of reinforcement learning. The three methods share the same structure in that the component weight vectors are represented by stochastic units drawn from the Gaussian distribution with mean $\mathbf{m}$ and variance $\beta^2 I$. We create adaptive update rules for these parameters of the Gaussian distribution so as to maximize the expected value of the long-term reward. We first derive a particular form of immediate reward reinforcement learning which can be applied to solve linear and non-linear projection problems. Then we suggest that the reinforcement learning neural network can be implemented for unsupervised projection problems. Lastly, based on temporal difference learning, we investigate two new algorithms that are based on Sarsa-learning and Q-learning for projection problems. We show that the last method has accurate convergence, even for non-linear projections.

Also, it is frequently important in projection methods to identify multiple components. Although we can find more than one component by deflationary methods such as the Gram-Schmidt method, these methods seem to be outwith the framework of reinforcement learning. Thus, we describe a general method that can find more components by re-defining the reward functions. To perform deflationary orthogonalization, we extend the definition of reward functions so that it includes one basic reward function and one extended reward function. Based on such an idea, two different ways are derived to identify orthogonal directions in linear projection problems and kernel methods.

8.1 Projection with Immediate Reward Learning

In this section, we apply immediate reward reinforcement learning to a set of projection methods. Under the framework of reinforcement learning, we consider that each agent is deemed to be taking actions in an environment

W.A. Barbakh et al.: Non-Stand. Param. Adapt. for Explor. Data Analys., SCI 249, pp. 123–149.
springerlink.com

that consists of data to be explored, in order to maximise the reward in this environment. Each agent has a set of parameters, $\mathbf{W}$, and these parameters are sampled by stochastic units drawn from $\mathcal{N}(\mathbf{m}, \beta^2\mathbf{I})$, the Gaussian distribution with mean $\mathbf{m}$ and variance β^2. Although different projection methods have their own objective functions, they share the common property that the reward function determines how well actions have been chosen according to the states in the environment. We thus show our method is quite a general one that can be easily applied to different projection methods.

8.1.1 An Example: Independent Component Analysis

As an example, we first consider the problem of independent component analysis. The goal of ICA is to recover independent signals given only a set of observations that are unknown linear mixtures of the independent signals. ICA is essentially based on higher-order information, while attempting to make the recovered signals as independent as possible.

The most common principle for ICA is to make all components as non-Gaussian as possible as described in Subsection 3.1.3. Thus, we optimize the demixing matrix $\mathbf{W}$ by measuring the kurtosis of the distribution of $\mathbf{Wx}$ where each component $\mathbf{w}_i$ of $\mathbf{W}$ corresponds to one stochastic unit. Given the prewhitened observations, we wish to maximize the absolute values of the kurtosis of each output component and the reward function is thus defined as

$$r_i = |\mathtt{kurt}(\mathbf{w}_i^T \mathbf{x})| \tag{8.1}$$

As we have stated in Subsection 6.2.1, the optimization of the reward function can be related to the update vector in weight space, $E\{\Delta\mathbf{W}|\mathbf{W}\}$. Then we have from (6.14),

$$\Delta w_{ij} = \alpha_{ij}(r_{i,\xi} - b_{ij}) \frac{\partial \ln P(y_i = \xi|\mathbf{w}_i, \mathbf{x})}{\partial w_{ij}}$$

We define that each component $\mathbf{w}_i$ is sampled from a Gaussian distribution, $\mathbf{w}_i \sim \mathcal{N}(\mathbf{m}_i, \beta_i^2)$ so that each component now has two parameters to learn, its mean $\mathbf{m}_i$ and the variance β_i^2 and we have

$$\log P(y_i = \xi|\mathbf{m}_i, \beta_i, \mathbf{x}) \propto -\frac{(\mathbf{w}_i - \mathbf{m}_i)^2}{\beta_i^2} - \frac{1}{\beta_i} \tag{8.2}$$

$$\Delta\mathbf{m}_i = \alpha_m(r_i - \bar{r}_i) \frac{\partial P(y_i = \xi|\mathbf{m}_i, \beta_i^2, \mathbf{x})}{\partial \mathbf{m}_i} \tag{8.3}$$

$$\Delta\beta_i = \alpha_\beta(r_i - \bar{r}_i) \frac{\partial P(y_i = \xi|\mathbf{m}_i, \beta_i^2, \mathbf{x})}{\partial \beta_i} \tag{8.4}$$

We thus update the parameters of the sampled distribution for each stochastic unit with the rules

$$\Delta \mathbf{m}_i = \alpha_m(r_i - \bar{r}_i)\frac{\mathbf{w}_i - \mathbf{m}_i}{\beta_i^2} \tag{8.5}$$

$$\Delta \beta_i = \alpha_\beta(r_i - \bar{r}_i)\frac{||\mathbf{w}_i - \mathbf{m}_i||^2 - \beta_i^2}{\beta_i^3} \tag{8.6}$$

where $\bar{r}$ is the average value of rewards received by this unit over all the iterations. Such an algorithm acts like a magnetic trap attracting all the greedy reward-seekers to identify the most kurtotic signal. We can see that the higher the reward function is, the more reward it will get, and so the value of the reward function will keep increasing provided the algorithm identifies the independent component with most kurtosis. However, given all the sets of weights being trained consistently, although the reward function can guarantee that all the recovered components are non-Gaussian, the algorithm tends to find the same original signal that has largest kurtosis. To avoid this issue, we perform the same reinforcement learning algorithm M times on different sets of weights in sequence and we orthogonalize the weight vectors $\mathbf{w}_1, \ldots, \mathbf{w}_M$ after each iteration using the Gram-Schmidt method of Subsection 3.1.5, i.e. we subtract the projections $(\mathbf{w}_p^T \mathbf{w}_j)\mathbf{w}_j, j = 1, \ldots, p-1$ from the current set of weights, $\mathbf{w}_p$ according to (3.56).

To illustrate our algorithm, we create a 2-dimensional artificial data set $\mathbf{s} = (\mathbf{s}_1, \mathbf{s}_2)$ of 1000 samples, where $\mathbf{s}_1 = \sinh(\mathbf{t}_1), \mathbf{s}_2 = \tanh(\mathbf{t}_2)$ and $\mathbf{t}_1, \mathbf{t}_2$ are vectors whose elements are from a Gaussian distribution, $\mathbf{t}_1, \mathbf{t}_2 \sim \mathcal{N}(0, 1)$. The 2×2 mixing matrix $\mathbf{A}$ is randomly generated and the mixed observations, $\mathbf{x} = \mathbf{A}\mathbf{s}$ are prewhitened to $\mathbf{z}$ as described in Subsection 3.1.3. We can see that with deflationary learning, our ICA algorithm has identified all the independent components in Figure 8.1. From Table 8.1, we can see that even if the difference in kurtosis between the original signals is large, the ICA algorithm can still identify ICs accurately.

Table 8.1. The kurtosis of the original signals, mixed observations and recovered ICs

	Kurtosis 1	**Kurtosis 2**
Original signals	16.4941	864.4390
Mixed observations	410.6288	31.5702
Recovered ICs	865.2676	16.4546

We can also evaluate the performance of our ICA algorithm using the Amari error [49], where we compare the final W and V, the true demixing matrix, by

$$d(V, W) = \frac{1}{2M}\sum_{i=1}^{M}\left(\frac{\sum_{j=1}^{M}|a_{ij}|}{\max_j |a_{ij}|} - 1\right) + \frac{1}{2M}\sum_{j=1}^{M}\left(\frac{\sum_{i=1}^{M}|a_{ij}|}{\max_i |a_{ij}|} - 1\right) \tag{8.7}$$

where $a_{ij} = (VW^{-1})_{ij}$ and V is the true demixing matrix. In this experiment, the Amari error is 0.0095.

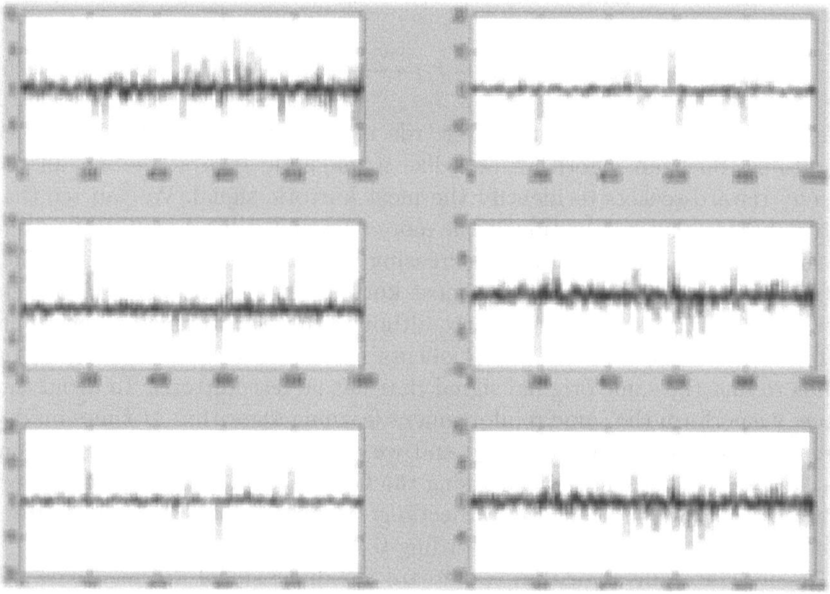

Fig. 8.1. Top: the original data set. Centre: the mixed observations. Bottom: the ICs recovered by our ICA algorithm.

Alternative Reward Functions

We derive a set of alternative reward functions based on negentropy, described in Subsection 3.1.3. We define the reward function by

$$r = ||G(y) - G(v)|| \tag{8.8}$$

where $G(y)$ is a nonquadratic function that does not grow fast and v is a standardized Gaussian random variable. The reward function means the greater the approximate negentropy is, the more reward it will get. We can also use a family of $G(y)$, defined in (3.24)-(3.28), to produce a family of reward functions.

To examine these reward functions, we use a 2-dimensional data set, 'chirp' and 'gong', provided by Matlab. The size of the data set is 1000. All the experiments have been performed with the same number of iterations. To identify different components, the Gram-Schmidt method is used. We can see that all of the reward functions can somewhat identify ICs as shown in Figure 8.2, but the performance with different reward functions is not the same, as shown in Table 8.2. [122] has stated that with measuring non-Gaussianity by negentropy, choosing a function $G(y)$ that is monotonic and does not grow too fast, one obtains more robust estimators. From the plot of the function curves of (3.24)-(3.28) in Figure 8.3, we can see the better the function $G(y)$

Table 8.2. Kurtosis by a family of reward functions based on (8.8)

	Kurtosis 1	**Kurtosis 2**
Original signals	3.2958	3.9589
Mixed observations	3.5871	3.8613
RF1	3.4228	3.9800
RF2	3.2961	3.9404
RF3	3.2865	3.9565
RF4	3.3647	3.9558
RF5	3.4549	3.8651

meets these criteria, the higher performance it has. Table 8.3 also reflects the same conclusion. We can also compare the results with that using the original reward function (8.1) in Table 8.3, which shows that reward function (8.1) is more efficient than the reward function (8.8).

Table 8.3. The Amari error for different $G(y)$, (3.24)-(3.28). **RF** corresponds to the Amari error with reward function (8.1).

	Amari error
RF	0.2150
RF1	0.5696
RF2	0.1066
RF3	0.0042
RF4	0.6516
RF5	0.2243

Non-linear PCA with immediate reward Learning

It [177] has been proved that there is a relationship between non-linear PCA and ICA in that given the observations have been whitened and thus the demixing matrix W is orthogonal, the criterion of non-linear PCA, $J(W) = E\{||\mathbf{x} - \mathbf{W}\mathbf{g}(\mathbf{W}^\mathbf{T}\mathbf{x})||^\mathbf{2}\}$ equals to the contrast function of ICA $J_{kurt}(W) = \sum_{i=1}^{n} E\{y_i^4\}$. Maximization/minimization of the criterion of non-linear PCA equals maximizing/minimizing the kurtosis of ICs. Therefore, we can combine the non-linear PCA algorithm with immediate reward reinforcement learning to separate independent components and define the reward function as

$$r = \frac{1}{1 + \exp(\gamma||\mathbf{x} - \mathbf{W}\tanh(\mathbf{W}^T\mathbf{x})||^2)}. \tag{8.9}$$

To examine the reward function (8.9), we use the same real data set as was used in the previous subsection and we can see that the ICA algorithm has separated the mixed observations into ICs correctly, as shown in Figure 8.4. The Amari error is 0.1598. Figure 8.4 shows the recovered components.

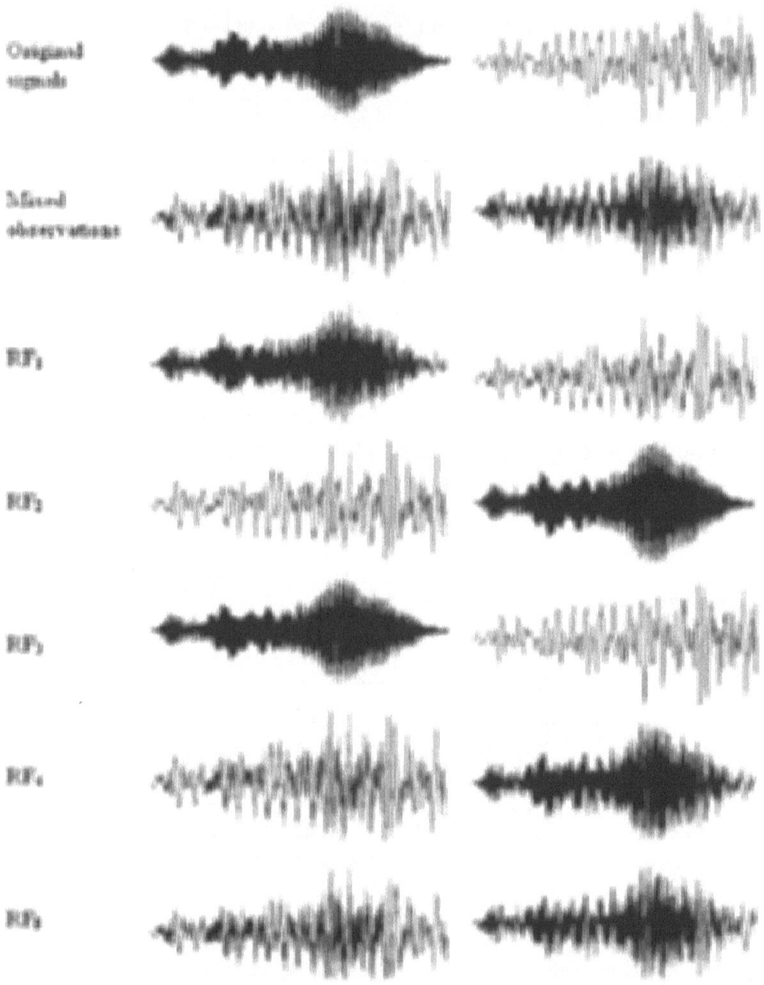

Fig. 8.2. The original signals, mixed observations and recovered ICs by the reward functions as labeled in the left column

Speech Simulation

As an example, we have combined the ICA algorithm with immediate reward reinforcement learning in former subsections in different ways. In this subsection, we use a 3-D real data set in which the original signals are pieces of records from different humans speaking, in which the number of observations is 10000. All of the records are collected at 8000Hz, 8bits and 1 channel, as shown in the top row of Figure 8.5. The real data set is mixed linearly as shown in the middle row of Figure 8.5, which partly simulates the situation of a cocktail party. The reason we say 'partly' here is that for a real cocktail

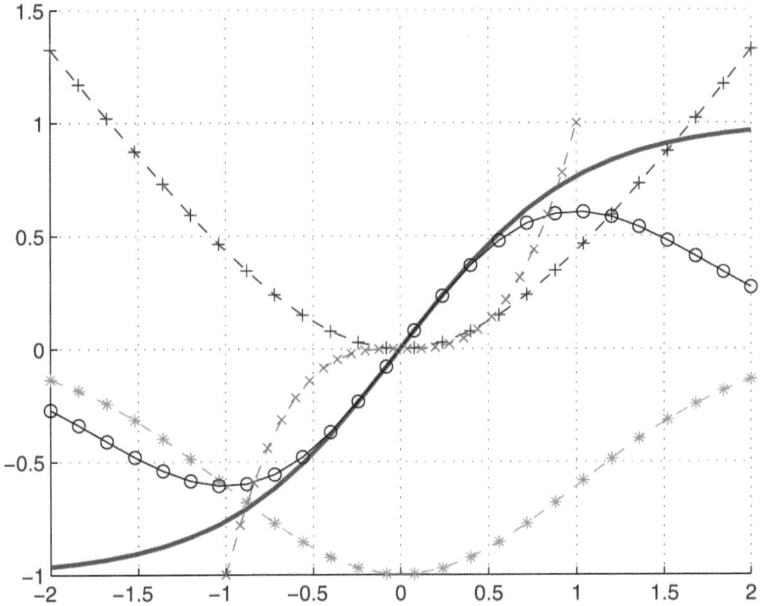

Fig. 8.3. Plot of the function curves of $G(y)$, (3.24)-(3.28). $G_1(y)$ in Eq. (3.24) is given by the dashed line with plus '+'. $G_2(y)$ in Eq. (3.25) is given by the dashed line with star '*'. $G_3(y)$ in Eq. (3.26) is given by the solid line. $G_4(y)$ in Eq. (3.27) is given by the solid line with 'o'. $G_5(y)$ in Eq. (3.28) is given by the dashed line with cross 'x'.

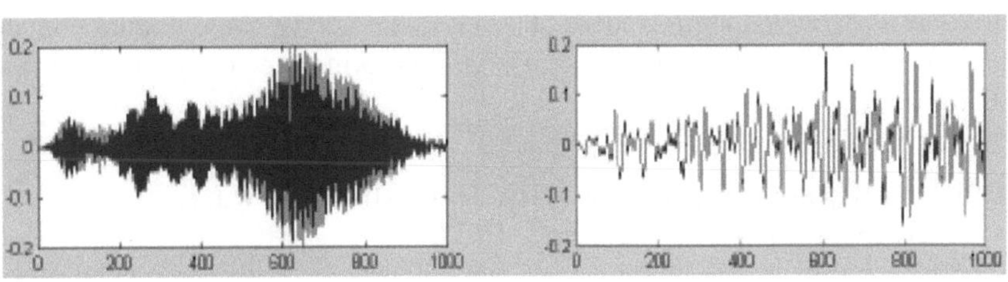

Fig. 8.4. ICs recovered by non-linear PCA with immediate reward reinforcement learning

party, the observations could be mixed non-linearly and may include reverberations. To identify different independent components, the Gram-Schmidt method is used.

It is clear that all independent components have been separated, as shown at the bottom of Figure 8.5. Empirically, we find that using real data usually requires a larger number of iterations than using artificial data and the

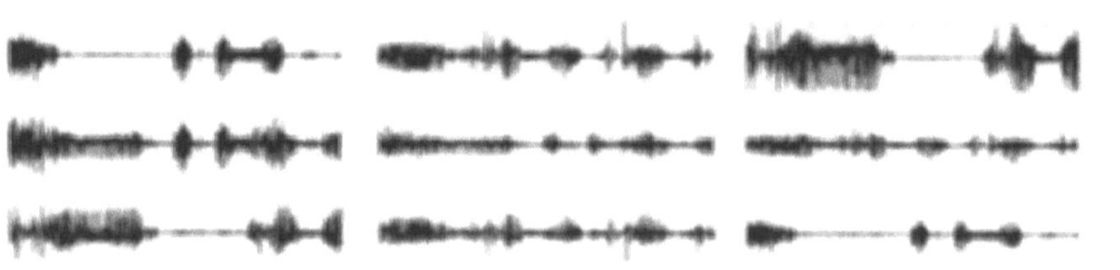

Fig. 8.5. Simulation with our ICA algorithm. Top: the original data set. Centre: the mixed observations. Bottom: the ICs recovered by reward function (8.1).

Table 8.4. Correlation between the original sources and recovered signals after 50000 iterations

0.0136	**0.9995**	0.0295
0.9989	0.0187	0.0419
0.0716	0.0118	**0.9974**

Table 8.5. Correlation between the original sources and recovered signals after 10000 iterations

0.0279	**0.9889**	0.1824
0.9929	0.1286	0.0633
0.0473	0.1414	**0.9817**

learning rate should be small so that learning processing can be performed smoothly and efficiently. Table 8.4 shows the correlation between the original sources and recovered signals after 50000 iterations. We show partial convergence in Table 8.5 after 10000 iterations. The Amari error for table 8.4 is 0.0695.

8.1.2 Multiple Components with Immediate Reward Reinforcement Learning - PCA

It is more frequently necessary in projection methods to identify multiple components. Principal component analysis finds the linear projection of a data set which contains maximal variance. For a centered data set, we define the reward function as $r = \frac{1}{1+\exp(-\gamma(\mathbf{w}^T\mathbf{x})^2)}$, where we take the sample weights $\mathbf{w}$, drawn from $\mathcal{N}(\mathbf{m}, \beta^2\mathbf{I})$ and multiply by the data sample at each iteration. We create a 5-dimensional artificial data set of 10000 samples, whose elements are drawn independently from Gaussian distribution with $\mathbf{x}_i \sim \mathcal{N}(0, i^2)$, where i is the index of the elements of the data set and $i = 1, 2, 3, 4, 5$. So $\mathbf{x}_5$ has the greatest variance and $\mathbf{x}_1$ has the lowest variance. With the Gram-Schmidt method in immediate reward learning, we can see that the five principal components have been identified clearly as shown in Table 8.6.

Table 8.6. The weights from the artificial data experiment for five principal components with Gram-Schmidt method

PC1	0.0214	-0.0514	-0.0194	0.0261	**-0.9979**
PC2	-0.0278	0.0566	0.0472	**0.9967**	0.0217
PC3	-0.0018	0.0741	**0.9957**	-0.0509	-0.0245
PC4	-0.0098	**-0.9943**	0.0778	0.0514	0.0509
PC5	**0.9930**	-0.0069	0.0043	0.0276	0.0224

One advantage that such learning has is that we can identify when the simulation is close to convergence by noting the value of the variance parameter: as we approach convergence, the variance tends towards 0 because all samples are very close to the optimal. We typically set a lower limit beyond which the variance cannot decrease. A reviewer has suggested that our methods would have some difficulty if the statistics of the data set did not allow unique principal components and so we have experimented with a data set, $\mathbf{x}_i \sim \mathcal{N}(0, i^2)$, where i is the index of the elements of the data set and $i = 1, 2, 4, 4, 5$. Thus there is an unique first principal component but the second variance forms a two dimensional plane within the 5 dimensional data. We first tried this with 200000 samples from this distribution but found that our algorithm is sensitive enough to identify the very small differences between the sample variances. Thus the second principal component filter was identified as (0,1,0,0,0) or (0,0,1,0,0). To investigate further we simply took samples from $\mathbf{x}_4 \sim \mathcal{N}(0, 16)$, and copied them into $\mathbf{x}_3$. Of course this now counts as a single dimension having variance =32 i.e. more that that which we allocated to the first direction. It is clear to see in table 8.7 that the two dimensions having the same variances together correspond to the first principal component direction and then the fifth dimension is identified to the second one and so on. We consider since the third and fourth dimension share the same variance, the algorithm considers these two dimensions simultaneously by taking sum of the variances in these two dimensions, which is 32 and is thus more than the fifth dimension, 25. We can see in table 8.8 that if we create another artificial data with $i = 1, 2, 4, 4, 6$, the fifth dimension has the largest variance and is identified as the first principal component direction. It is worth noting that since the algorithm considers the two dimensions with the same variance simultaneously by taking sum of the variances, there are actually four PC directions to be identified. Thus, after estimating the first four PC directions, one direction being the same as the first PC filter but in opposite direction is identified to be the fifth PC direction.

However such a deflationary method seems to be outwith the framework of reinforcement learning itself, because $\Delta\mathbf{m}$ is also partly decided by the deflationary method instead of only being decided by the reward function. Therefore we derive a new method in which deflationary orthogonalization is directly combined into the reward function so that the change in Gaussian parameters is only decided by the reward function. The basic idea is that

Table 8.7. The weights from the artificial data experiment for five principal components with Gram-Schmidt method, where there are two elements in the data set exactly having the same variance, $i = 1, 2, 4, 4, 5$

PC1	0.0841	0.0041	-0.0323	**0.9952**	-0.0386
PC2	0.0378	0.0134	**-0.9986**	-0.0357	-0.0028
PC3	**0.7084**	0.1038	0.0274	-0.0324	**0.6968**
PC4	**0.6860**	0.0910	0.0322	-0.0851	**-0.7162**
PC5	-0.1382	**0.9903**	0.0078	0.0076	-0.0070

Table 8.8. The weights from the artificial data experiment for five principal components with Gram-Schmidt method, where there are two elements in the data set exactly having the same variance, $i = 1, 2, 4, 4, 6$

PC1	0.0706	0.0209	-0.0074	**0.9953**	-0.0627
PC2	-0.0021	0.0007	**-0.9999**	-0.0066	-0.0100
PC3	0.0252	**0.7134**	-0.0061	-0.0572	**-0.6980**
PC4	0.0207	**0.7005**	0.0073	0.0288	**0.7127**
PC5	**0.9970**	0.0020	-0.0019	-0.0725	-0.0280

the angle between the current weight vector $\mathbf{w}_j$ and $\mathbf{w}_k, k = 1, \ldots, j-1$ previously estimated is one factor to consider when updating parameters. Since we want the current weight vector $\mathbf{w}_j$ to be orthogonal to other weight vectors $\mathbf{w}_k, k = 1, \ldots, j-1$, the immediate reward is weakened by eliminating sum of angle between the current weight vector $\mathbf{w}_j$ and $\mathbf{w}_k, k = 1, \ldots, j-1$, $\sum_{k=1}^{j-1} \mathbf{w}_j^T \mathbf{w}_k$ or sum of projection on $\mathbf{w}_k, k = 1, \ldots, j-1$, $\sum_{k=1}^{j-1} (\mathbf{w}_j^T \mathbf{w}_k)\mathbf{w}_k$. Obviously, we need not to do so for the first principal component, and we re-define the reward function as

$$r = \begin{cases} \frac{1}{1+\exp(-\gamma(\mathbf{w}_j^T \mathbf{x}_i)^2)}, & j = 1 \\ \frac{1}{1+\exp(-\gamma(\mathbf{w}_j^T \mathbf{x}_i)^2 + \sum_{k=1}^{j-1} (\mathbf{w}_j^T \mathbf{w}_k)^n)}, & j = 2, \ldots, K \end{cases} \tag{8.10}$$

or sum of projection

$$r = \begin{cases} \frac{1}{1+\exp(-\gamma(\mathbf{w}_j^T \mathbf{x}_i)^2)}, & j = 1 \\ \frac{1}{1+\exp(-\gamma(\mathbf{w}_j^T \mathbf{x}_i)^2 + \sum_{k=1}^{j-1} [(\mathbf{w}_j^T \mathbf{w}_k \mathbf{w}_k)^T (\mathbf{w}_j^T \mathbf{w}_k \mathbf{w}_k)]^n)}, & j = 2, \ldots, K \end{cases} \tag{8.11}$$

where the data is K-dimensional and i is the index of the data point. We need to point out that the reason we define the parameter n is that the reward function becomes unstable if the value of $\gamma(\mathbf{w}_j^T \mathbf{x})^2$ and the value of $\sum_{k=1}^{j-1} (\mathbf{w}_j^T \mathbf{w}_k)\mathbf{w}_k$ are too close to each other. We believe this is because it is more difficult to generate the correct immediate reward in such a situation. At each iteration, we specifically normalise each $\mathbf{w}$ so that its length is equal to 1.

Thus with the deflation incorporated into the reward function directly, we need to first form a reward function for the first component according to a particular projection method and then extend this reward function in order to identify more than one component by (8.10) or (8.11). Through experiments, we found (8.10) is more efficient than (8.11). We can apply this idea in the following chapters in this book and we denote the reward function for the first component as the *Basic Reward Function* and the reward function for the subsequent components as the *Extended Reward Function*. Thus the reward function includes two parts: the basic reward function for the first component and extended reward functions for subsequent components.

Table 8.9. The weights from the artificial data experiment for five principal components by re-defining the reward function as (8.11)

PC1	-0.0088	0.0195	0.0153	-0.0035	**-0.9990**
PC2	-0.0134	0.0718	-0.0559	**-0.9945**	-0.0076
PC3	0.0127	0.0208	**0.9986**	-0.0515	-0.0106
PC4	-0.0130	**0.9963**	-0.0021	0.0914	-0.0222
PC5	**0.9934**	0.0372	0.1387	0.0021	0.0370

We test our method with the same artificial data set used above. We choose (8.11) as the reward function, where $\gamma = 1, n = 3$. The number of iterations for each weight vector is 200,000 and the learning rate is 0.1, which decreases linearly to zero. Table 8.9 shows that all the principal components have been found and we can see the weight vectors converge smoothly as shown in Figure 8.6.

8.1.3 Simulation: Canonical Correlation Analysis

We continue with an example simulation showing this method used for classical Canonical Correlation Analysis (CCA), in which we attempt to find the linear combination of the variables that gives us maximum correlation between the combinations. We can regard such a problem as that of maximizing the objective function $g_1(\mathbf{w}_1|\mathbf{w}_2) = E(y_1 y_2)$. However such problem is an unconstrained maximization problem which clearly has no finite solution. To constrain the maximization, we add the constraint that $E(y_1^2 = 1)$ and similar with y_2. We define a set of reward functions, where r_i corresponds to the reward on the ith data set, $i = 1, 2$. Based on the same idea as in the previous subsection, we add another factor to the reward function to constrain the maximization and also we want to find multiple canonical correlations, which yields the reward function for classic CCA,

$$r_{ij} = \begin{cases} \frac{1}{1+\exp(\gamma||\mathbf{w}_{1j}^T\mathbf{x}_1 - \mathbf{w}_{2j}^T\mathbf{x}_2||^2 + |(\mathbf{w}_{ij}^T\mathbf{x}_i)^2 - 1|)}, & j = 1 \\ \frac{1}{1+\exp(\gamma||\mathbf{w}_{1j}^T\mathbf{x}_1 - \mathbf{w}_{2j}^T\mathbf{x}_2||^2 + |(\mathbf{w}_{ij}^T\mathbf{x}_i)^2 - 1| + \sum_{k=1}^{j-1}(\mathbf{w}_{ij}^T\mathbf{w}_{ik})^n)}, & j = 2, \ldots, K \end{cases} \quad (8.12)$$

where i is the index of the data set, j is the index of the correlation component we want to estimate and k is the index of correlation components we have previously estimated.

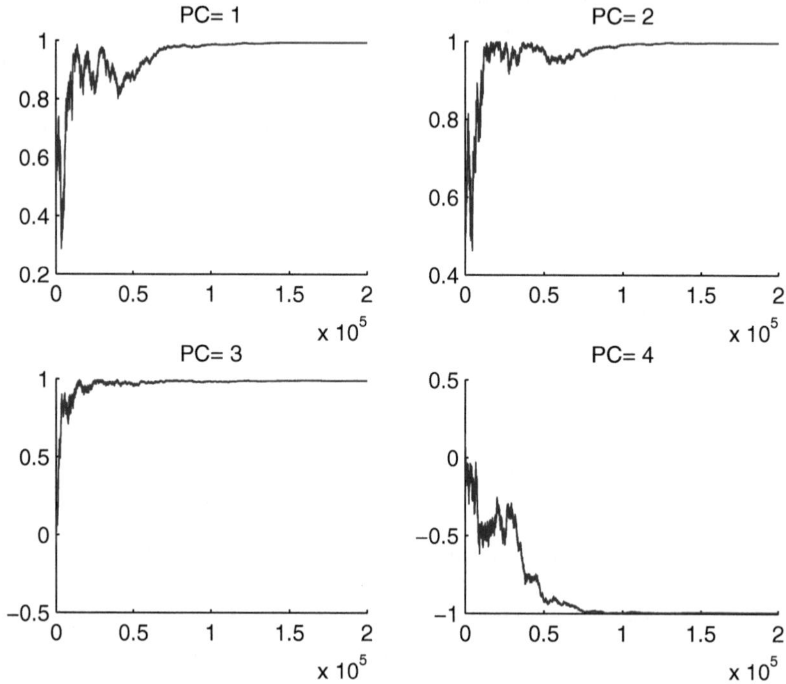

Fig. 8.6. Convergence of the PCA weight vectors to the optimal directions. For all such graphs, the vertical axis shows the cosine of the angle between the current filter and the optimal filter. The horizontal axis shows the number of iterations.

We use an artificial data set similar to what has been used in [143], in which there are two sets of artificial data, one is 4-dimensional and the other is 3-dimensional: each of the elements is drawn from the zero-mean Gaussian distribution, $\mathcal{N}(0,1)$ and we add an additional sample from $\mathcal{N}(0,1)$ to the first elements of each vector and then divide by 2 to ensure that there is no more variance in the first elements than in the others. To generate the second correlation, we add an additional sample from $\mathcal{N}(0,0.5)$ to the second elements of each vector, so that the second correlation is smaller than the first one. We use the reward function (8.12), where $i = 2, K = 2, \gamma = 1, n = 3$. We set the initial learning rate to 0.1, which is decreased linearly to zero over 200000 iterations.

We illustrate the convergence of the CCA weight vectors in Figure 8.7: we can see consistent convergence towards the first canonical correlation vector smoothly and then towards the second canonical correlation vector, which is slightly smoother than the first one. Table 8.10 shows that the reward function works well in identifying the different canonical correlations.

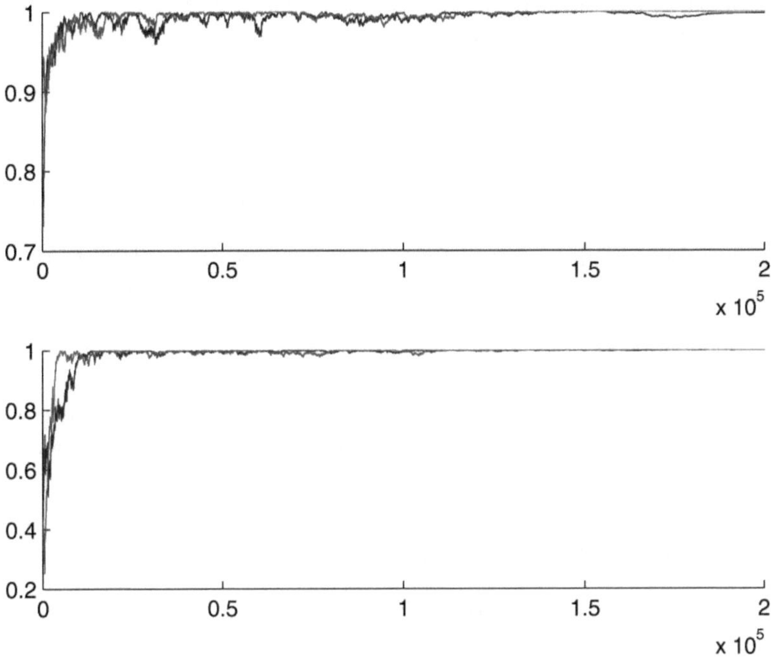

Fig. 8.7. Convergence of the CCA weight vectors to the optimal directions. Top: the $\mathbf{1^{st}}$ canonical correlation. Bottom: the $\mathbf{2^{nd}}$ canonical correlation.

Table 8.10. The first two Canonical Correlation (CC) weight vectors found with the artificial data

	m1	m2
1st CC	**0.9992**	**0.9995**
	0.0173	0.0009
	-0.0263	-0.0316
	-0.0250	
2nd CC	-0.0028	-0.0306
	0.9986	**0.9993**
	-0.0509	0.0216
	0.0102	

8.1.4 Deflationary Orthogonalization for Kernel Methods – Kernel PCA

We illustrate deflationary orthogonalization in the reward function for kernel methods with Kernel principal component analysis (KPCA) as an example. KPCA searches for a set of weight vectors in the feature space that maximizes the variance of the projections. As usual for kernel methods, one direction $\mathbf{w}$ is parameterized by a $N \times 1$ weight vector $\alpha = (\alpha_1, \ldots, \alpha_N)^T$ such that

$$\mathbf{w} = \alpha^T \Phi_x = \sum_{j=1}^{N} \alpha_j \Phi(\mathbf{x}_j) \in \mathcal{F} \tag{8.13}$$

Considering KPCA as a kernel eigenvalue problem and α is the eigenvector of the scalar product matrix, K, the first principal component direction in the feature space can be found by finding $\alpha = \arg\max_\beta \beta^T K^T K \beta$ and we follow the same argument as in [198] to identify the first principal component direction by finding

$$\alpha = \arg\max_\beta \beta^T K \beta. \tag{8.14}$$

For more than one principal component in feature space, we wish to find directions in the feature space such that they are orthogonal to each other, so that

$$\mathbf{w}_i^T \mathbf{w}_j = 0 \tag{8.15}$$

where i, j are the index of directions. We have

$$\mathbf{w}_i^T \mathbf{w}_j = \alpha_i^T \Phi_{\mathbf{x}} \Phi_{\mathbf{x}}^T \alpha_j = \alpha_i^T K \alpha_j = 0. \tag{8.16}$$

Therefore, the problem reduces to finding the weight vectors $\alpha_j, j = 1, \ldots, d$ that satisfies (8.14) and (8.16). For the first principal component in the feature space, we define the reward function as

$$r = \alpha_1^T K \alpha_1 \tag{8.17}$$

To discover multiple components, we extend this reward function so that it satisfies (8.16). Assuming we have found $j-1$ directions by immediate reward reinforcement learning, to ensure the current direction $\mathbf{w}_j$ is orthogonal to the other $j-1$ directions, we need to subtract the sum of projections between the current weight vector α_j and the weight vectors $\alpha_k, k = 1, \ldots, j-1$ in the feature space, so the current weight vector becomes $\alpha_j^{'} = \alpha_j - \sum_{k=1}^{j-1} (\alpha_j^T K \alpha_k) \alpha_k$. We denote $\Sigma_\alpha = \sum_{k=1}^{j-1} (\alpha_j^T K \alpha_k) \alpha_k$, for $j \geq 2$, and we have

$$\begin{aligned} r &= (\alpha_j^{'})^T K \alpha_j^{'} \\ &= (\alpha_j - \Sigma_\alpha)^T K (\alpha_j - \Sigma_\alpha) \\ &= \alpha_j^T K \alpha_j - \alpha_j^T K \Sigma_\alpha - \Sigma_\alpha^T K \alpha_j + \Sigma_\alpha^T K \Sigma_\alpha \end{aligned} \tag{8.18}$$

In practice, we can omit $\Sigma_\alpha^T K \Sigma_\alpha$ and we re-define the reward function as

$$r_j = \begin{cases} \alpha_j^T K \alpha_j, & j = 1 \\ \alpha_j^T K \alpha_j - \alpha_j^T K \Sigma_\alpha - \Sigma_\alpha^T K \alpha_j, & j > 1 \end{cases} \tag{8.19}$$

where j is the index of direction we are searching for.

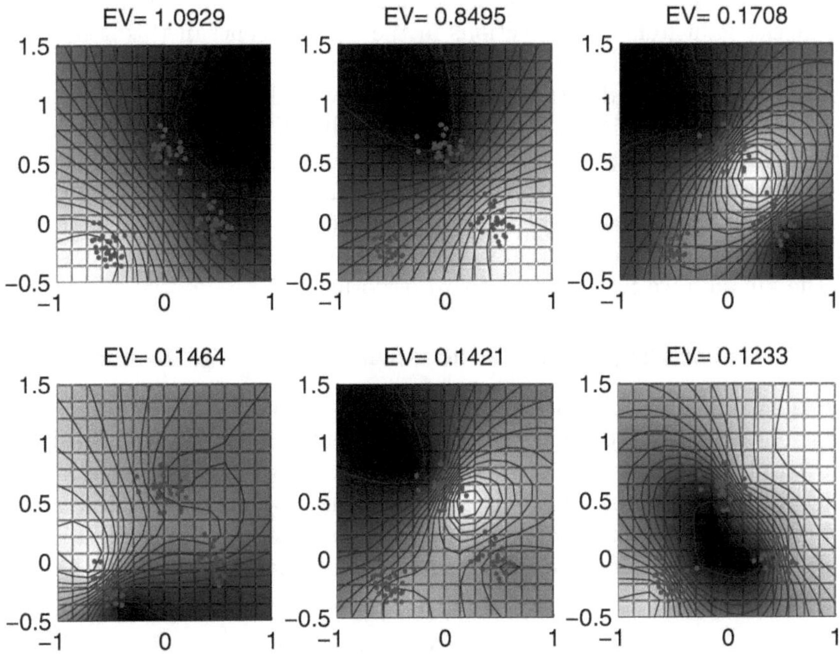

Fig. 8.8. The first 6 directions of principal components for a 2-dimensional 3-cluster data set by immediate reward reinforcement learning. Each contour line represents constant principal component value.

To illustrate reward function (8.19) for KPCA, we create a 2-dimensional data set of 90 samples, in which the first $\frac{1}{3}$ are centered at $(0.5, 0.2)$, the next $\frac{1}{3}$ at $(0, 0.6)$ and the last at $(0.5, 0)$. We use an exponential radial basis kernel so $K_{ij} = \exp(-\gamma||\mathbf{x}_i - \mathbf{x}_j||)$. We centre the kernel matrix by using $\widetilde{K} = K - \frac{1}{N}\mathbf{1}^T K - \frac{1}{N}K\mathbf{1} + \frac{1}{N}\mathbf{1}^T K\mathbf{1}$. We use the reward function (8.19) and $\gamma = 0.1$ in the kernel function. The initial learning rate is 0.1, which is decreased linearly to zero over 10000 iterations. The result is shown in Figure 8.8 and we see that the first 6 directions are similar to those found by the standard methods.

8.2 Projections with Stochastic Synapses

In the previous section, we have demonstrated projection methods in immediate reward reinforcement learning. We see that reinforcement learning describes a group of techniques for parameter adaptation based on a methodology which envisages an agent making an exploratory investigation of its environment with a view to identifying the optimal strategy for actions within the environment where optimal is defined in terms of the reward which an agent can gain from taking actions in the environment. In this section, we review a second form of immediate reward reinforcement learning in which the individual unit is deterministic but has stochastic synapses. Four learning rules have been developed from this perspective in Subsection 6.2.2 and we investigate the use of these learning rules to perform linear and non-linear projection techniques such as principal component analysis, exploratory projection pursuit, canonical correlation analysis and kernel methods. The method is quite a general one and simply requires a reward function which is specific to the function we require the unit to perform.

8.2.1 Linear Projection Methods with Stochastic Weights

We follow the same structure for reinforcement learning as in Section 8.1: each agent is deemed to be taking actions in the environment (which consists of the data to be explored) in order to maximise his reward in this environment. Each agent is represented by one individual unit which is deterministic but has stochastic weights sampled from the Gaussian distribution. [155] has noted that the weight change gradient can be estimated as $E\{\Delta\mathbf{w}|\mathbf{w}\}$, which satisfies

$$\begin{aligned} E\{\Delta\mathbf{w}|\mathbf{w}\} &= \eta\sum_{\mathbf{x}} r(\mathbf{x})\mathrm{e}(\mathbf{x})p(\mathbf{x},\mathbf{w}) \\ &= \eta E\{r\mathrm{e}|\mathbf{w}\} \\ &= \eta\nabla_{\mathbf{w}} E\{r|\mathbf{w}\}. \end{aligned} \tag{8.20}$$

where $e()$ as before is the score function.

Therefore, in order to increase the average reward we could optimize the stochastic weights $\mathbf{w}$ iteratively, where the four learning rules stated in Subsection 6.2.2 and listed in Table 8.11 could be used. We illustrate the method with Principal Component Analysis (PCA) and then discuss the reward functions for other linear projections.

Table 8.11. Learning rules A1, A2, B1 and B2 from Subsection 6.2.2

Learning Rule	Formula
Rule A1	$\Delta\mu_{ij} = \eta r(y_i - \mu_{y_i})x_j$
Rule A2	$\Delta\mu_{ij} = \eta[r(y_i - \mu_{y_i}) + \lambda(1-r)(-y_i - \mu_{y_i})]x_j$
Rule B1	$\Delta\mu_{ij} = \eta r(w_{ij} - \mu_{ij})$
Rule B2	$\Delta\mu_{ij} = \eta[r(w_{ij} - \mu_{ij}) + \lambda(1-r)(-w_{ij} - \mu_{ij})]$

Principal Component Analysis

Principal Component Analysis (PCA) finds the linear filter onto which projections of a data set have greatest variance. Thus the agent is rewarded in proportion to its variance. Let the agent currently select actions from $\mathcal{N}(\mu_t, \sigma_t^2)$

and let $\mathbf{w}_t$ be the action which the agent has selected at time instant t. For centered data, $\mathbf{x} \in X$, we can simply take $r = (\mathbf{w}_t^T \mathbf{x})^2$ for principal component analysis or the more complex but equally effective $r = \frac{1}{1+\exp(-\gamma(\mathbf{w}^T\mathbf{x})^2)}$. In Figure 8.9, we illustrate convergence with the first of these on an artificial data set in which each element of $\mathbf{x}, x_i, i = 1, \ldots, 5$ is drawn from $\mathcal{N}(0, i^2)$ i.e. the first principal component filter is $(0,0,0,0,1)$. Note that in (6.28), $y = \mathbf{w}_t^T \mathbf{x}$ while $\mu_{y_i} = \mu_t^T \mathbf{x}$.

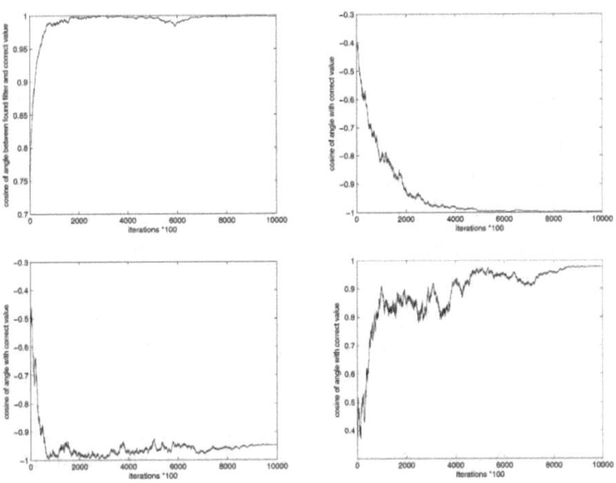

Fig. 8.9. Convergence of the 1st principal component weight vector with stochastic synapses. Top left: convergence with rule A1. Top Right:convergence with rule A2. Bottom left: convergence with rule B1. Bottom Right: convergence with rule B2.

The method clearly finds the first principal component in all cases, however the A rules seem to be more efficient than the B rules. Also, considering the solution for PCA is in a quadratic surface without local minimum, it is not necessary to improve A1 rule and B1 rule by adding the small anti-trapping λ-term, which gives the A2 rule and B2 rule. Results show that the A1 rule and B1 rule seem to be more efficient than the A2 rule and B2 rule respectively. Furthermore, we are often interested in finding more than the first principal component of a data set. This may be easily done by a similar method to that of Subsection 8.1.2. Thus we have

$$r_k = (\mathbf{w}_k^T \mathbf{x})^2 - \sum_{j=1}^{k-1} (\mathbf{w}_k^T \mu_j \mu_j)^T (\mathbf{w}_k^T \mu_j \mu_j) \tag{8.21}$$

where we have used the subscript k to denote the kth filter. Results from a similar data set as before and using the rule A2 are shown in Table 8.12. Again we see all principal components have been found with great accuracy.

Table 8.12. Each column corresponds to one of the principal component filters found by the method. The principal components have been found in order of magnitude of the variance of their projections.

PC1	-0.0152	0.0212	0.0747	0.0502	**0.9947**
PC2	0.0012	-0.0370	-0.0192	**0.9952**	0.0209
PC3	0.0222	0.0393	**0.9932**	0.0174	-0.0399
PC4	-0.0305	**0.9974**	-0.0341	-0.0073	-0.0397
PC5	**-0.9992**	0.0418	-0.0803	-0.0819	0.0839

Other linear projections

This method is quite general and we can use the method to perform:

- Minor Component Analysis which finds the filter with least variance which has been used in regression and in spectral clustering. The reward function is then $r = \exp(-(\mathbf{w}_t\mathbf{x})^2)$. By using the same artificial data set as in the previous subsection, exemplar convergence for the first minor component is shown in Figure 8.10.
- Canonical Correlation Analysis (CCA): this method searches for the linear filter within two data sets simultaneously which maximizes the correlations between the respective projections. Thus we may define the basic reward function in form of $r = \frac{1}{1+\exp(\gamma(\mathbf{w}_1^T\mathbf{x}_1-\mu_1^T\mathbf{x}_1)(\mathbf{w}_2^T\mathbf{x}_2-\mu_2^T\mathbf{x}_2))}$ and to identify multiple correlations, we can re-define the basic reward function as (8.12) in Subsection 8.1.3. We test our algorithm with the same artificial data set used in Subsection 8.1.3. The result is shown in Figure 8.11 and we can see that the algorithm has identified the two canonical correlations clearly as shown in Table 8.13.

Table 8.13. The first two Canonical Correlation (CC) weight vectors found with the artificial data

	m1	**m2**
1st CC	**-0.9981**	**-0.9936**
	0.0015	-0.0637
	-0.0384	0.0937
	0.0482	
2nd CC	-0.0173	0.0792
	-0.9961	**-0.9900**
	0.0055	0.1169
	0.0862	

8.2.2 Kernel Methods with Stochastic Weights

As described in Subsection 3.2.1, kernel methods are a recent addition to the data analysts tool package. With the same use of deflationary orthogonalization for kernel methods as in Subsection 8.1.4, the basic reward function for

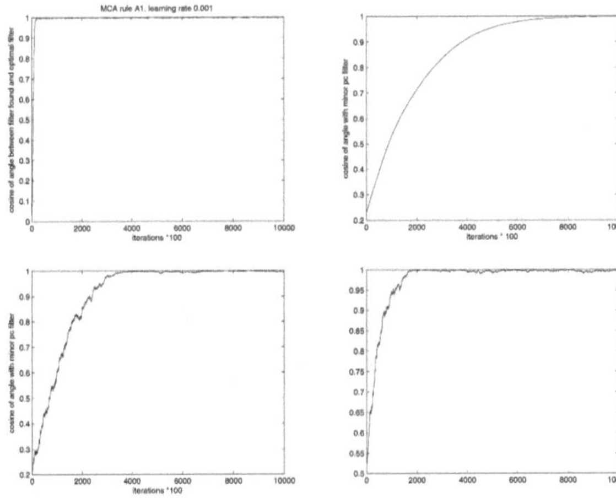

Fig. 8.10. Convergence of the 1st minor component weight vector with stochastic synapses. Top left: convergence with rule A1. Top Right:convergence with rule A2. Bottom left: convergence with rule B1. Bottom Right: convergence with rule B2.

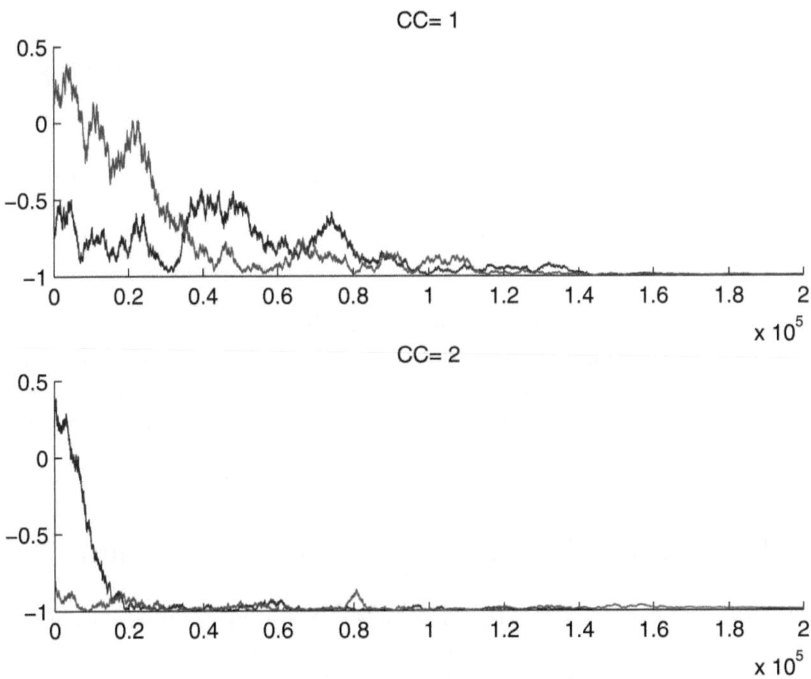

Fig. 8.11. Convergence of the CCA weight vectors to the optimal directions. Top: the 1st canonical correlation. Bottom: the 2nd canonical correlation.

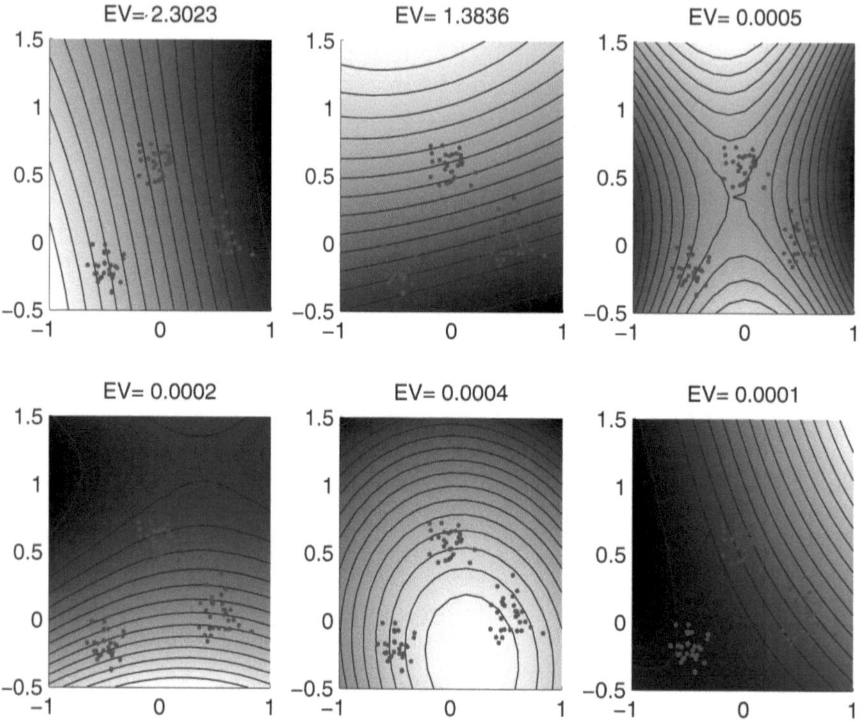

Fig. 8.12. The first 6 directions of kernel principal components for the 2-dimensional 3-cluster data set used in Subsection 8.1.4. Each contour line represents constant principal component value.

KPCA is defined as $r = \mathbf{w}^T K \mathbf{w}$, where K is the kernel matrix. We test our algorithm with the same artificial data set used in Subsection 8.1.4, and we use a radial basis kernel so $K_{ij} = \exp(-\gamma||\mathbf{x}_i - \mathbf{x}_j||^2)$. Figure 8.12 shows the first six principal components in feature space identified by this method with A1 rule.

8.3 Projection with Temporal Difference Learning

The temporal difference (TD) learning methods in reinforcement learning algorithms have been applied to a variety of tasks, such as dispatch problems and robot navigation. In this section, we present two similar methods for projection problems with the two main forms of temporal difference learning, Sarsa-learning and Q-learning. We will apply the first method with Q-learning to the linear projection problems and then apply the second method with Sarsa-learning to kernel projection problems. We will demonstrate that both of our methods can converge to the optimal solution and especially, the method with Sarsa-learning gives more stable and accurate results than those using immediate reward learning.

8.3.1 Linear Projection with Q-Learning

In this subsection, we derive a method to solve linear projection problems with the Q-learning method. We illustrate the method with principal component analysis.

The Q-learning method has been introduced in Subsection 6.2.3. This method directly approximates the optimal action-value function, Q^*, by the learned action-value function, Q, and the best possible action selected in the subsequent state:

$$Q(s_t, a_t) \leftarrow Q(s_t, a_t) + \alpha[r_{t+1} + \gamma \max_a Q(s_{t+1}, a) - Q(s_t, a_t)].$$

We therefore structure the learning algorithms: the state of the system at any time is the data sample presented to the system at that time, i.e. we equate s_t with $\mathbf{x}_t$, $s_t = \mathbf{x}_t$. We use a parametric (actually Gaussian) estimator for the statistic to be calculated and the action taken in response to the state is to sample the weight vector $\mathbf{w}$ from the distribution $\mathcal{N}(\mathbf{m}, \beta^2 I)$ with the current estimate of the parameters, $\mathbf{m}$ and β^2, having been optimized in the previous iteration. Then we may identify rewards and update both the estimated Q-value (which is what we wish to maximise) of the estimator and the parameters (mean and variance) of the estimator.

We use the same 5 dimensional data as before in which the first principal component is readily identified as the fifth input dimension. At each stage, we keep a note of which data point, say the i^*th, has the greatest Q-value; note that we are only interested in a Q-value for each state/data point since we only have one action each time. Then when we generate our sample, $\mathbf{w}$, from the Gaussian distribution, $\mathcal{N}(\mathbf{m}, \beta^2 I)$, we calculate one of the following

$$\Delta Q_i \leftarrow \alpha(r + \gamma Q_{i^*} - Q_i) \tag{8.22}$$

$$\Delta Q_i = \alpha(r + \gamma \mathbf{m}^T \mathbf{x}_{i*} - Q_i) \tag{8.23}$$

where, in each case, $r = (\mathbf{w}^T \mathbf{x}_i)^2$. Notice that we have labelled the current data point as $\mathbf{x}_i$ while the data point which currently has the greatest Q-value is labelled $\mathbf{x}_{i*}$. In each iteration, the calculation is then followed by

$$Q_i \leftarrow Q_i + \Delta Q_i \tag{8.24}$$

$$\mathbf{m}_j = \mathbf{m}_j + \eta * \Delta Q_i * (\mathbf{w}_j - \mathbf{m}_j), j = 1, ..., D \tag{8.25}$$

where i is the index of data point we have randomly selected, D is the dimensionality of the data and η is the learning rate.

The number of iterations is 200000 and the learning rate is initialized to 0.1 and reduced linearly to zero. Figure 8.13 shows the results of the methods in equations (8.22) and (8.23). Although both of the methods have identified that the fifth input dimension corresponds to the first principal component, the former method gives more accurate convergence than the latter.

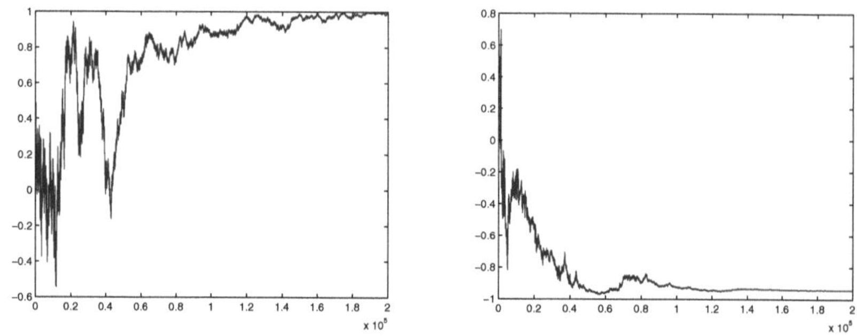

Fig. 8.13. The convergence of $\mathbf{m}$ for the two methods in equation (8.22) (left) and equation (8.23) (right) respecively. The former seems to give more accurate convergence than the latter.

8.3.2 Non-linear Projection with Sarsa Learning

In this subsection, we derive a second way to find non-linear projections with sarsa learning. We illustrate the method with kernel PCA and kernel CCA.

Sarsa learning has been introduced in Subsection 6.2.3. This method computes the expected return based on the state-action pairs which gives the state-action value function for the current policy π. The sarsa algorithm involves transitions from state-action pair to state-action pair, and learns the value of state-action pairs using Eq. (6.38),

$$Q(s_t, a_t) \leftarrow Q(s_t, a_t) + \alpha[r_{t+1} + \gamma Q(s_{t+1}, a_{t+1}) - Q(s_t, a_t)].$$

In the previous Subsection 8.3.1, the state of the system is defined as the data sample presented to the system at that time, i.e. $s_t = \mathbf{x}_t$. With sarsa learning, we still use the Gaussian estimator to sample the weight vector $\mathbf{w}$ from the distribution $\mathcal{N}(\mathbf{m}, \beta^2 I)$ with the current estimate of the parameters, the mean $\mathbf{m}$ and the variance β^2. We now view the current parameters $\mathbf{m}^{(t)}$ and β^2 at that time to be the state. The action taken in response to the state is to sample a weight vector $\mathbf{w}^{(t)}$ from the distribution $\mathcal{N}(\mathbf{m}^{(t)}, \beta^2 I)$ with the current parameters $\mathbf{m}^{(t)}$ and β^2 having been optimized in the previous iteration. Then we may identify rewards and update both the estimated state-action value (which is what we wish to maximise) of the estimator and the parameters (mean and variance) of the estimator.

In accordance with (6.38), the state-action value, $Q(s_{t+1}, a_{t+1})$, needs to be estimated at each iteration t in order to update the current estimated state-action value. We keep in mind that since we have defined the states to lie in a continuous space, the state-action value function will also be continuous. Therefore, we estimate the state-action value by forming a state-action value function $Q(s, a, \theta)$ parameterized by parameter θ, whose input is the state-action pair. A function approximation method is applied to optimize the parameters θ in the state-action value function.

Consider the problem of kernel principal component analysis. Again, we want to search for a set of weight vectors α in the feature space that maximizes the variance of the projections. At iteration t, the weight vector $\alpha^{(t)}$ is drawn from the Gaussian distribution $\mathcal{N}(\mathbf{m}^{(t)}, \beta^2\mathbf{I})$, where $\mathbf{m}^{(t)}$ is $N \times 1$ vector and in practice, the value of β is kept at 1. After $\alpha^{(t)}$ has been drawn from $\mathcal{N}(\mathbf{m}^{(t)}, \beta^2\mathbf{I})$, we have the immediate reward r_t according to (8.19). The state-action value function will be approximated by two-layer multi-layer perceptrons with recursive least squares so that the state-action value, $Q(s_{t+1}, a_{t+1})$, can be estimated through the current optimized parameter $\theta^{(t)}$ and the current state-action pair $\alpha^{(t)}$. We then calculate the following

$$\Delta Q_t = \alpha(r_t + \gamma Q(s_{t+1}, a_{t+1}) - Q(s_t, a_t)) \tag{8.26}$$

$$Q(s_t, a_t) \leftarrow Q(s_t, a_t) + \Delta Q_t \tag{8.27}$$

$$\mathbf{m}_j = \mathbf{m}_j + \eta * \Delta Q_t * (\alpha_j - \mathbf{m}_j), j = 1, ..., D \tag{8.28}$$

where D is the dimension of principal components we attempt to indentify in feature space and $\alpha, \mathbf{m}$ are $N \times 1$ vectors. The parameters of the multi-layer perceptrons with recursive least squares which we have used to approximate the state-action value function, $\theta^{(t)}$, are finally optimized with the input $\alpha^{(t-1)}$ and the target value, $Q(s_t, a_t)$ calculated by (8.27).

With the algorithm above and adapting our deflation method of Subsection 8.1.4, we use the same data set as in Subsection 8.1.4 and the initial learning rate is still 0.1, which is decreased linearly to zero over 10000 iterations. We plot only the first two principal component weight vectors as shown in Figure 8.14 and we see that the TD method has identified different

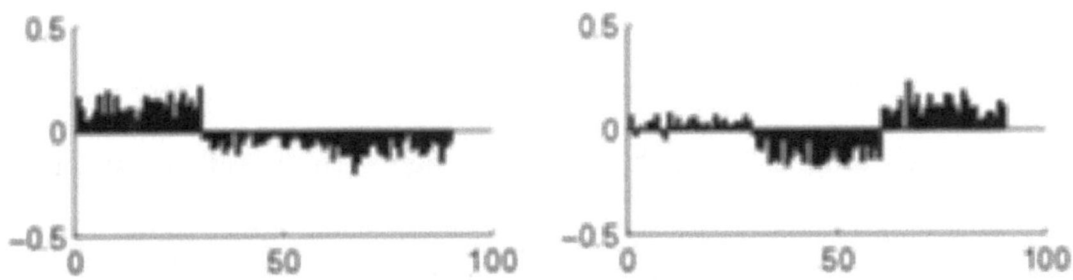

Fig. 8.14. Plot of the first two principal component vectors in feature space by reinforcement learning with sarsa-learning. Left: Plot of the weight vector in the first principal component direction. Right: Plot of the weight vector in the second principal component direction. We see that our TD method has identified different clusters of the data set more clearly.

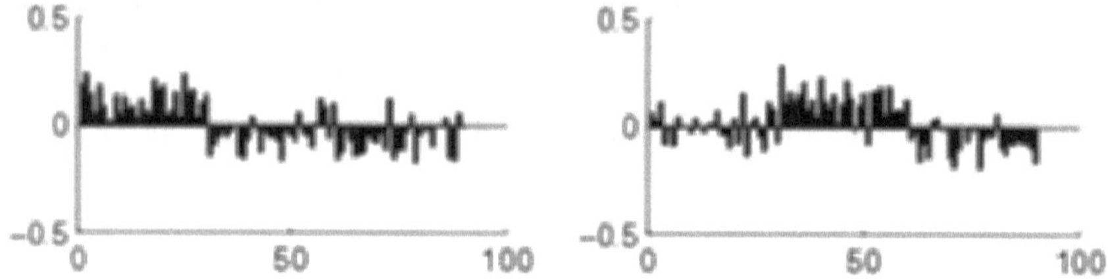

Fig. 8.15. Plot of the first two principal component vectors in feature space by immediate reward reinforcement learning of Subsection 8.1.4. Left: Plot of the weight vector in the first principal component direction. Right: Plot of the weight vector in the second principal component direction.

clusters of the data set very clearly, which is much better than the results shown in Figure 8.15 by the algorithm in Section 8.1.4.

A simulation: Kernel CCA

Considering the problem of kernel canonical correlation analysis, for the first canonical correlation in feature space, the basic reward function is denoted as

$$\begin{aligned} r_\alpha &= \alpha^T K_1 K_2 \beta - \gamma|\alpha^T K_1 K_1^T \alpha - 1| \\ r_\beta &= \alpha^T K_1 K_2 \beta - \gamma|\beta^T K_2 K_2^T \beta - 1| \end{aligned} \qquad (8.29)$$

Based on the same idea as in Subsection 8.1.4, we can extend the reward function (8.29) to support more than one canonical correlation in feature space, so that the canonical correlation directions of each data set are orthogonal to each other. Thus by denoting $\Sigma_\alpha = \sum_{k=1}^{j-1}(\alpha_j^T K_1 \alpha_k)\alpha_k$ and $\Sigma_\beta = \sum_{k=1}^{j-1}(\beta_j^T K_2 \beta_k)\beta_k$, for $j \geq 2$, we have

$$\begin{aligned} r_0 &= (\alpha_j^{'})^T K_1 K_2 \beta_j^{'} \\ &= (\alpha_j - \Sigma_\alpha)^T K_1 K_2 (\beta_j - \Sigma_\beta) \qquad (8.30) \\ &= \alpha_j^T K_1 K_2 \beta_j - \alpha_j^T K_1 K_2 \Sigma_\beta - \Sigma_\alpha^T K_1 K_2 \beta_j + \Sigma_\alpha^T K_1 K_2 \Sigma_\beta \qquad (8.31) \end{aligned}$$

In practice, we omit $\Sigma_\alpha^T K_1 K_2 \Sigma_\beta$ and we re-define the reward function, for $j \geq 2$, as

$$\begin{aligned} r_{\alpha_j} &= \alpha_j^T K_1 K_2 \beta - \alpha_j^T K_1 K_2 \Sigma_\beta - \Sigma_\alpha^T K_1 K_2 \beta_j - \gamma|\alpha_j^T K_1 K_1^T \alpha_j - 1| \\ r_{\beta_j} &= \alpha_j^T K_1 K_2 \beta - \alpha_j^T K_1 K_2 \Sigma_\beta - \Sigma_\alpha^T K_1 K_2 \beta_j - \gamma|\beta_j^T K_2 K_2^T \beta_j - 1| \end{aligned} \qquad (8.32)$$

where j is the index of direction we are searching for.

We apply the same temporal difference learning algorithm to Kernel CCA with the reward function (8.32). Since each reward corresponds to each of the data sets, we use two agents with the same structure. For the first agent, the weight vector $\alpha^{(t)}$ is drawn from the Gaussian distribution $\mathcal{N}(\mathbf{m_1}^{(t)}, \sigma^2\mathbf{I})$, where $\mathbf{m_1}^{(t)}$ is $N \times 1$ vector and in practice, the value of the variance σ is kept at 1. We sample the weight vector β in the second agent in the same way. Then we have the immediate reward r_α and r_β according to (8.32). The value function for the first agent is approximated with the function approximation method with parameters $\theta_{\mathbf{1}}^{(t)}$. The second agent is defined in the same way with the parameters $\theta_{\mathbf{2}}^{(t)}$.

We want to investigate multiple canonical correlations in feature space and so we create two 2-dimensional data sets of 150 samples. For the first data set, we centre the first $\frac{1}{3}$ data at $(1.5, 1.5)$, the next at $\frac{1}{3}$ at $(-1.5, 1.5)$ and the last $\frac{1}{3}$ is drawn from a uniform distribution in the interval $(-1.5, 1.5) \times (1.5, 1.5)$. The second data set has the same structure, which the first $\frac{1}{3}$ data is centered at $(-1.5, -1.5)$, the next $\frac{1}{3}$ at $(1.5, -1.5)$ and the last $\frac{1}{3}$ is drawn from a uniform distribution in the interval $(-1.5, -1.5) \times (1.5, -1.5)$. Thus, both of

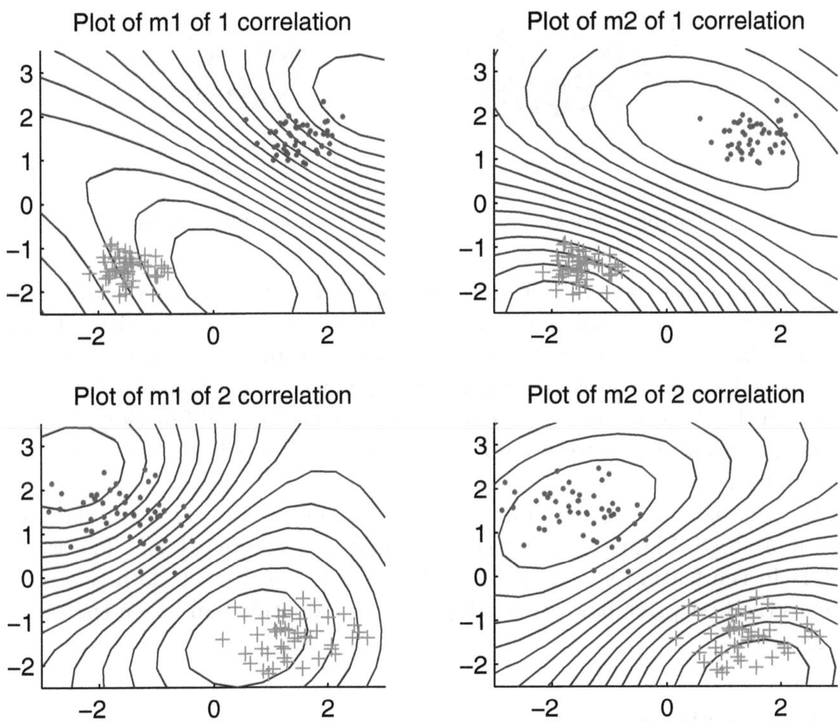

Fig. 8.16. The first 2 directions of principal relationships. Top: the first kernel canonical correlation; Bottom: the second kernel canonical correlation. The lines show lines of equal correlations. The dots represent the first data set and the '+' represent the second data set.

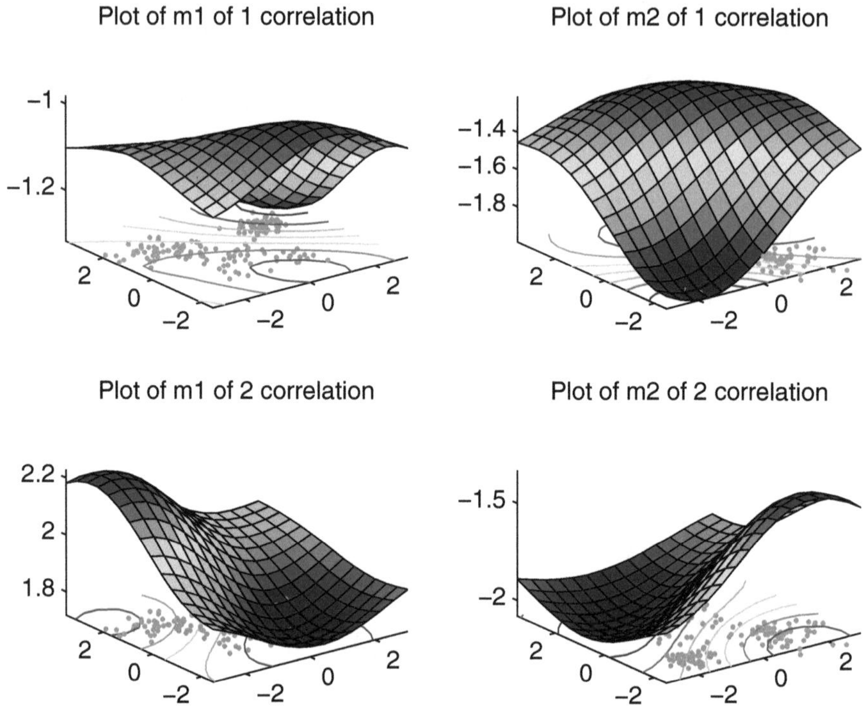

Fig. 8.17. The first 2 directions of principal relationships with sarsa-learning

the two data sets have two clusters. We smear the first nonlinear correlation by adding an additional noise vector from $\mathcal{N}(0, 0.3\mathbf{I_2})$ to the first clusters of each data set. Then the second noise vector from $\mathcal{N}(0, 0.6\mathbf{I_2})$ is added to the second clusters of each data set, so that we have a weaker relationship between the data points in this cluster. We use the reward function of (8.32). The initial learning rate is 0.001, which is decreased linearly to zero over 30000 iterations.

We see that our algorithm has identified different clusters in the data sets in Figure 8.16. Figure 8.17 shows the results in a different form. The weight filters in the feature space focus on the first cluster in each data set, which means the first principal relationship has been identified and then the second is found.

8.4 Conclusion

We have, in this chapter, utilised three reinforcement learning methods based on different learning strategies. We first present a particular form of immediate reward reinforcement learning in which the reward for an action is immediate. We define each agent by a set of parameters, **W** sampled by stochastic units drawn from $\mathcal{N}(\mathbf{m}, \beta^2\mathbf{I})$, the Gaussian distribution with mean $\mathbf{m}$ and variance β^2. We incorporate the objective of different projection problems into the reward function to evaluate how well actions have been chosen according to the states in the environment, and then the parameters are optimized by updating mean $\mathbf{m}$ and variance β^2 of the Gaussian distribution iterately. We first demonstrated our immediate reward reinforcement learning on independent component analysis. A set of reward functions has been developed from different perspectives and the results are clear. The reinforcement learning method is such a general method that it can easily be applied to other projection problems. We have used the method to perform linear projections and kernel projections.

We have also combined deflationary orthogonalization directly into the reward function so that the action is only decided by the reward function. The basic idea is to make sure all the component weight vectors are orthogonal to each other and we do so in two different ways for linear projection methods and kernel methods respectively. We draw the conclusion that the reward function can be extended to contain multiple factors with different meanings so that the reinforcement learning is constrained under more than one condition. A good example is the simulation of canonical correlation analysis, whose reward function is not only defined to identify the canonical correlations, but also with the factor that limits the CCA to have a finite solution and the factor that performs deflationary orthogonalization.

We have developed alternative immediate reward reinforcement learning in which the individual unit has stochastic synapses. We have investigated four learning rules for unsupervised projection problems. The results shows that the method can converge smoothly with these learning rules but they have different efficiencies.

Meanwhile, since immediate reward reinforcement learning does not allocate incremental rewards to actions which is one of the most powerful features of reinforcement learning, it suffers from the drawback of converging slowly and may be unstable. We have, finally, used two algorithms of reinforcement learning based on temporal difference learning, one of which is based on Q-learning and the other of which is based on Sarsa-learning. We first applied the algorithm with Q-learning to principal component analysis. We have applied the algorithm with Sarsa-learning to kernel PCA and kernel CCA and compared the results with those from immediate reward learning. The TD method has the advantages of converging more quickly with more accurate and stable results. Additionally, for kernel projections, we have also developed a different way to identify multiple directions in the feature space and shown clear and compelling results.

9

Cross Entropy Methods

9.1 The Cross Entropy Method

The cross-entropy (CE) method [56] is a general Monte Carlo approach to combinatorial and continuous multi-extremal optimization [195]. The method originated from the field of rare event simulation, where very small probabilities need to be accurately estimated.

Consider a Monte Carlo simulation which draws instances from the true distribution of events. Such a problem usually requires an inordinate number of draws before enough of the rare events are seen to make a reliable estimate of their probability of occurring. A better way is to use *Importance Sampling* (IS) [7, 66, 118] which is a general technique for estimating the properties of a particular distribution, while only having samples generated from a different distribution than the distribution of interest. However, such a method has the drawback that the optimal reference parameters to be used in IS are usually difficult to obtain. The cross entropy method thus has the advantage that it provides a simple adaptive procedure for estimating the optimal reference parameters with asymptotic convergence properties. Not only being applied to the estimates of *rare-event* problems in dynamic models [57, 140], a simple modification of the CE method [195] has been also applied to solve difficult combinatorial optimization problems (COPs), see ([193, 194]).

9.1.1 Rare-Event Simulation via Cross Entropy

Generally speaking, the basic mechanism of cross entropy involves an iterative procedure of two phases:

1. draw random data samples from the currently specified distribution.
2. identify those samples which are, in some way, "closest" to the rare event of interest and update the parameters of the currently specified distribution to make these samples more representative in the next iteration.

Let $l = (S(\mathbf{x}) > \gamma)$ be the event in which we are interested and typically we will be interested in problems in which l is very small. We could use Monte Carlo methods to estimate l but if l is very small this would lead to a very

W.A. Barbakh et al.: Non-Stand. Param. Adapt. for Explor. Data Analys., SCI 249, pp. 151–174.
springerlink.com

large number of samples before we could get reliable estimates of l. The cross entropy method uses *importance sampling* rather than simple Monte Carlo methods: if the original pdf of the data is $f(\mathbf{x})$, then we require to find a pdf, $g(\mathbf{x})$, such that all of $g()$'s probability mass is allocated in regions in which the samples are close to the rare-event. More formally, we have the deterministic estimate

$$l = \int I_{\{S(\mathbf{x})>\gamma\}} f(\mathbf{x}) d\mathbf{x} = \int I_{\{S(\mathbf{x})>\gamma\}} \frac{f(\mathbf{x})}{g(\mathbf{x})} g(\mathbf{x}) d\mathbf{x} = E_{g()} \left[I_{\{S(\mathbf{X})>\gamma\}} \frac{f(\mathbf{X})}{g(\mathbf{X})} \right]. \tag{9.1}$$

where I_L is the indicator function describing when L in fact occurred. An unbiased estimator of this is

$$\hat{l} = \frac{1}{N} \sum_{i=1}^{N} I_{\{S(\mathbf{X_i})>\gamma\}} \frac{f(\mathbf{x}_i)}{g(\mathbf{x}_i)} = \frac{1}{N} \sum_{i=1}^{N} I_{\{S(\mathbf{X_i})>\gamma\}} \mathbf{W}(f(\mathbf{x}_i), g(\mathbf{x}_i)) \tag{9.2}$$

where $\mathbf{W}()$ is known as the likelihood ratio.

The best $g()$ in (9.1) is $g^*(\mathbf{x}) = \frac{I_{\{S(\mathbf{x})>\gamma\}} f(\mathbf{x})}{l}$, which would have the same shape as $f()$ but all its probability mass in the interesting region. This is illustrated in Figure 9.1. Note that for the optimal $g()$, $\int_{\mathbf{x}:S(\mathbf{x})>\gamma} g*(\mathbf{x}) d\mathbf{x} = 1$ while $\int_{\mathbf{x}:S(\mathbf{x})>\gamma} f(\mathbf{x}) d\mathbf{x} = l$.

However we don't know l. (This is exactly what we are trying to get). So what we do is to pick a family of PDFs $g(\mathbf{x}, \mathbf{v})$, parametorised by $\mathbf{v}$ ($\mathbf{v}$ will be the mean and variance for Gaussian).

We now wish to minimise Kullback Leibler divergence between g^* and $g()$,

$$\min KL(g^*, g) = \int g^*(\mathbf{x}) \ln g^*(\mathbf{x}) d\mathbf{x} - \int g^*(\mathbf{x}) \ln g(\mathbf{x}, \mathbf{v}) d\mathbf{x} \tag{9.3}$$

So we maximise the cross entropy $\int g^*(\mathbf{x}) \ln g(\mathbf{x}, \mathbf{v}) d\mathbf{x}$.

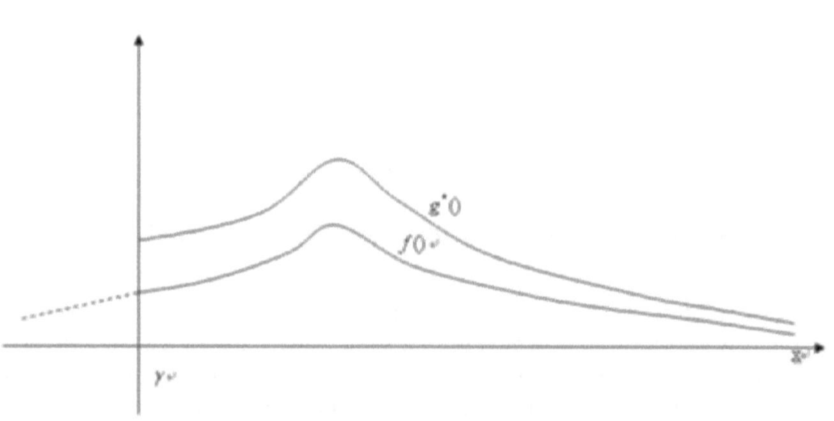

Fig. 9.1. The original distribution f() has probability mass outwith the region we are interested in but the importance sampling distribution has only domain $\mathbf{x} : S(\mathbf{x}) > \gamma$

We pick $\mathbf{v}$ to $\max \int \frac{I_{\{S(\mathbf{x})>\gamma\}} f(\mathbf{x})}{l} \ln g(\mathbf{x}, \mathbf{v}) d\mathbf{x}$, which is the same as

$$\max \int I_{\{S(\mathbf{x})>\gamma\}} f(\mathbf{x}) \ln g(\mathbf{x}, \mathbf{v}) d\mathbf{x}, \tag{9.4}$$

where we have discarded l a constant. But getting an optimal $g(\mathbf{x}, \mathbf{v})$ for a particular γ may not be an easy task. Therefore we create a set of γ_t for which we estimate the corresponding $\mathbf{v}_t$. The γ_t are chosen such that

$$P(\mathbf{x} : S(\mathbf{x}) > \gamma_t) > P(\mathbf{x} : S(\mathbf{x}) > \gamma_{t+1}) \tag{9.5}$$

i.e. at each iteration, the events are becoming more rare under $f()$. Therefore

$$\max \int I_{\{S(\mathbf{x})>\gamma\}} f(\mathbf{x}) \ln g(\mathbf{x}, \mathbf{v}) d\mathbf{x} \tag{9.6}$$

$$= \max_{\mathbf{v}_t} \int I_{\{S(\mathbf{x})>\gamma_t\}} \frac{f(\mathbf{x})}{g(\mathbf{x}, \mathbf{v}_{t-1})} \ln g(\mathbf{x}, \mathbf{v}_t) g(\mathbf{x}, \mathbf{v}_{t-1}) d\mathbf{x} \tag{9.7}$$

$$= \max_{\mathbf{v}_t} E_{g(\mathbf{x},\mathbf{v}_{t-1})} \left\{ I_{\{S(\mathbf{x})>\gamma_t\}} \mathbf{W}(f(\mathbf{x}), g(\mathbf{x}, \mathbf{v}_{t-1})) \ln g(\mathbf{x}, \mathbf{v}_t) \right\} \tag{9.8}$$

This is deterministic but we are working with samples. So we pick $\mathbf{v}_t$ to maximise

$$\max_{\mathbf{v}_t} \frac{1}{N} \sum_{i=1}^{N} I_{\{S(\mathbf{x}_i)>\gamma_t\}} \mathbf{W}(f(\mathbf{x}_i), g(\mathbf{x}_i, \mathbf{v}_{t-1})) \ln g(\mathbf{x}, \mathbf{v}_t) \tag{9.9}$$

For example, if $g(\mathbf{x}, \mathbf{v}) = \frac{1}{\sqrt{2\pi\sigma^2}} e^{-\frac{1}{2}(\frac{\mathbf{x}-\mu}{\sigma})^2}$, we find minimum of

$$\frac{1}{N} \sum_{i=1}^{N} I_{\{S(\mathbf{x}_i)>\gamma_t\}} \mathbf{W}_{t-1} \left\{ \ln(\sigma) + \frac{1}{2\sigma^2} (\mathbf{x} - \mu)^2 \right\} \tag{9.10}$$

We calculate the derivative of this with respect to the parameters, and set this equal to 0, to determine

$$\hat{\mu} = \frac{\sum_{i=1}^{N} I_{\{S(\mathbf{X}_i)>\hat{\gamma}_t\}} W(\mathbf{X}_i, \mathbf{u}, \hat{\mathbf{v}}_{t-1}) X_i}{\sum_{i=1}^{N} I_{\{S(\mathbf{X}_i)>\hat{\gamma}_t\}} W(\mathbf{X}_i, \mathbf{u}, \hat{\mathbf{v}}_{t-1})} \tag{9.11}$$

$$\hat{\sigma^2} = \frac{\sum_{i=1}^{N} I_{\{S(\mathbf{X}_i)>\hat{\gamma}_t\}} W(\mathbf{X}_i, \mathbf{u}, \hat{\mathbf{v}}_{t-1}) (\mathbf{x}_i - \hat{\mu})^2}{\sum_{i=1}^{N} I_{\{S(\mathbf{X}_i)>\hat{\gamma}_t\}} W(\mathbf{X}_i, \mathbf{u}, \hat{\mathbf{v}}_{t-1})} \tag{9.12}$$

Then the simplest algorithm [56] depends on working within a family of pdfs whose parameters we update i.e. let $f(\mathbf{x}) = f(\mathbf{x}, \mathbf{u})$, $\mathbf{u}$ being a parameter of the family to which $f()$ belongs; then the basic algorithm is

1. Define $\hat{\mathbf{v}}_0 = \mathbf{u}$. Set t=1.
2. Generate random samples, $\mathbf{X}_1, ..., \mathbf{X}_N$ from $f(\mathbf{x}, \hat{\mathbf{v}}_{t-1})$.

3. Calculate $S(\mathbf{X}_1), ..., S(\mathbf{X}_N)$ and order them. Let $\hat{\gamma}_t$ be the $1-\rho$ sample quantile, above which we identify the "elite" samples.
4. Use the same samples to calculate

$$\hat{\mathbf{v}}_t = \frac{\sum_{i=1}^{N} I_{\{S(\mathbf{X}_i) > \hat{\gamma}_t\}} W(\mathbf{X}_i, \mathbf{u}, \hat{\mathbf{v}}_{t-1}) h(\mathbf{x}_i)}{\sum_{i=1}^{N} I_{\{S(\mathbf{X}_i) > \hat{\gamma}_t\}} W(\mathbf{X}_i, \mathbf{u}, \hat{\mathbf{v}}_{\mathbf{t-1}})} \tag{9.13}$$

   where $h(\mathbf{X}_i)$ is capturing some statistic of the elite samples: for example, if we have a Gaussian random variable, h() would be defining the mean and variance of the distribution.
5. If $\hat{\gamma}_t = \gamma$, continue; else $t = t + 1$ and return to 2
6. Generate a sample $\mathbf{X}_1, ..., \mathbf{X}_{N_1}$ from $f(\mathbf{x}, \hat{\mathbf{v}}_t)$ and estimate

$$l = \frac{1}{N_1} \sum_{i=1}^{N_1} I_{\{S(\mathbf{X}_i) > \hat{\gamma}_t\}} W(\mathbf{X}_i, \mathbf{u}, \hat{\mathbf{v}}_{t-1}) \tag{9.14}$$

Note that, although step 4 looks formidable, it is actually only counting the fraction of samples which satisfy the current criterion. For fixed $\hat{\gamma}_t$ and $\hat{\mathbf{v}}_{t-1}$, we derive $\hat{\mathbf{v}}_t$ in step 4 with the following target

$$\max_{\mathbf{v}} \hat{D}(\mathbf{v}) = \max_{\mathbf{v}} \frac{1}{\mathbf{N}} \sum_{i=1}^{N} I_{\{S(\mathbf{X}_i) \geq \hat{\gamma}_t\}} W(\mathbf{X}_i, \mathbf{u}, \hat{\mathbf{v}}_{t-1}) \ln f(\mathbf{X}_i; \mathbf{v}). \tag{9.15}$$

The "cross entropy method" is so-called since we wish to minimise the Kullback-Leibler divergence between the data distribution and the importance sampling distribution which is related to the cross entropy via

$$CE_{f(.,v)}(f(.,u)) = H(f(.,v)) + KL(f(.,v), f(.,u)) \tag{9.16}$$

where CE is the cross entropy, KL is the Kullback-Leibler divergence and $H(.)$ is the Shannon entropy. Since $H(.)$ is constant, [140] equates the cross entropy with the Kullback-Leibler divergence.

9.1.2 Combinatorial Optimization via Cross Entropy

We may also use the CE method for Combinatorial Optimization Problems (COPs). The main idea is to first turn each COP into a rare-event estimation problem, the so-called *associated stochastic problem*(ASP) and then perform the cross entropy method. We usually wish to maximize some performance function $S(\mathbf{x})$ over all states $\mathbf{x}$ in data set $\aleph$. Denoting the maximum by γ^*, we have

$$\gamma^* = \max_{\mathbf{x} \in \aleph} S(\mathbf{x}) \tag{9.17}$$

Thus, by defining a family of pdfs $\{f(.;\mathbf{v}), \mathbf{v} \in \mathcal{V}\}$ on the data set $\aleph$, we follow [140] to associate with (9.17) the following estimation problem

$$l(\gamma) = \mathbf{P_u}(S(\mathbf{X}) \geq \gamma) = \mathbf{E}_u I_{\{S(\mathbf{x}) > \gamma\}} \tag{9.18}$$

where $\mathbf{X}$ is a random vector with pdf $f(.\,;\mathbf{u}), \mathbf{u} \in \mathcal{V}$. Thus $\mathbf{P_u}(S(\mathbf{X})$ is the probability associated with the distribution parameterised by $\mathbf{u}$. To estimate l for a certain γ close to γ^*, we can make adaptive changes to the probability density function according to the Kullback-Leibler cross-entropy. Thus we create a sequence $f(.\,;\mathbf{u}), f(.\,;\mathbf{v_1}), f(.\,;\mathbf{v_2}), \ldots$ of pdfs that are optimized in the direction of the optimal density. By associating the underlying combinatorial optimization problem, that is to optimize S, with a rare-event probability of $\mathbf{P}_u(S(\mathbf{X}) \geq \gamma)$, we can obtain an estimate of the "reference parameter vector" $\mathbf{v}^*$ via the CE algorithm.

Similar to the algorithm described in the previous subsection, the process is as follows:

- For a fixed $\mathbf{v}_{t-1}$, we choose the $(1-\varrho)$-quantile of $S(\mathbf{X})$, γ_t, which satisfies

$$\mathbf{P}_{\mathbf{v}_{t-1}}(S(\mathbf{X}) \geq \gamma_t) \geq \varrho, \tag{9.19}$$

  The estimator $\hat{\gamma}_t$ can be obtained by drawing a set of random samples $\mathbf{X}_1, \ldots, \mathbf{X}_N$ and evaluating the sample $(1-\varrho)$-quantile of performances

$$\hat{\gamma}_t = S_{[(1-\varrho)N]} \tag{9.20}$$

- For fixed $\hat{\gamma}_t$ and $\hat{\mathbf{v}}_{t-1}$, we optimize $\hat{\mathbf{v}}_t$ by

$$\max_{\mathbf{v}} \hat{D}(\mathbf{v}) = \max_{\mathbf{v}} \frac{1}{\mathbf{N}} \sum_{i=1}^{N} I_{\{S(\mathbf{X}_i) \geq \hat{\gamma}_t\}} \ln f(\mathbf{X}_i; \mathbf{v}). \tag{9.21}$$

It is important to observe that in the ASP, given fixed $\hat{\gamma}_t$ and $\hat{\mathbf{v}}_{t-1}$, (9.21) is different from (9.15) in that (9.21) does not contain the likelihood ratio term W. This is because the initial reference parameter vector $\mathbf{u}$ is arbitrary, however using this parameter is essential when solving a rare event problem.

We have introduced the cross entropy method in Section 9.1: the cross entropy method was motivated as an adaptive algorithm for estimating probabilities of rare events in complex stochastic networks. In this chapter, we will first investigate several unsupervised projection methods with the cross-entropy method.

We start with independent component analysis with a batch cross entropy method. Then we derive a new on-line cross-entropy algorithm for linear projection problems, in which the parameters of the stochastic units are updated more smoothly. By using a variety of appropriate performance functions for projection problems, we show that all the results are accurate and stable. Although Gram-Schmidt deflation works well in identifying multiple components, in this chapter we will apply a similar method to that in the previous chapter to perform deflationary orthogonalization in the performance function directly.

We then extend the Maximum Likelihood method and the EM algorithm with latent variable models based on cross entropy methods. The projections are performed based upon probability models. We then perform the optimization of latent variable models as associated stochastic problems with batch or on-line cross entropy methods. The results illustrate that the method can identify the global optimum.

We finish this chapter by investigating deep architectures in unsupervised data exploration. We extend the topology preserving mapping with cross entropy to develop a multilayer topology preserving mapping for clustering. We demonstrate that different clusters can be identified more clearly in higher layers.

9.2 ICA as Associated Stochastic Problem

In Section 9.1, we stated that the main idea of combinatorial optimization via cross entropy methods is to first turn each COP into a rare-event estimation problem, the so-called *associated stochastic problem*(ASP). We wish to maximize some performance function $S(\mathbf{x})$ over all states $\mathbf{x}$ in data set $\aleph$. We define a family of pdfs $\{f(.\,;\mathbf{v}), \mathbf{v} \in \mathcal{V}\}$ on the data set and optimize the parameters $\mathbf{v} \in \mathcal{V}$ of these pdfs in the direction of the optimal density so that we can then obtain an estimate of the "reference parameter vector" $\mathbf{v}^*$ via the CE algorithm.

In this section, we investigate a new ICA algorithm with the cross-entropy method for optimization. We use stochastic units drawn from a Gaussian distribution to sample the elements of the demixing matrix, $\mathbf{w}_i$. With the CE algorithm, we make adaptive changes to the probability density function of $\mathbf{w}_i$ according to the value of the performance function. Since we adopt the basic idea that all components should be as non-Gaussian as possible, the performance function can be the measurement of the nongaussianity of a vector, for example by using kurtosis. The reference parameters of the pdfs are thus steered in the direction of the theoretically optimal density.

To be specific, we consider the same ICA problem as in Subsection 3.1.3: we wish to transform a set of observations $\mathbf{x}_1, \dots, \mathbf{x}_n$ that are the result of a linear mixing of statistically independent sources $\mathbf{s}_1, \dots, \mathbf{s}_n$ by $\mathbf{x} = \mathbf{A}\mathbf{s}$, into several components that are statistically independent by $\mathbf{y} = \mathbf{W}\mathbf{x}$. To maximize the absolute values of the kurtosis of each output component with the CE algorithm for optimization, we wish to maximize the performance function

$$S_i(\mathbf{w}_i) = |\mathtt{kurt}(\mathbf{w}_i^T \mathbf{x})| \tag{9.22}$$

where i is the index of the independent component. We denote the maximum by γ^*, thus $\gamma^* = \max_{\mathbf{w}} S(\mathbf{w})$. We consider the associated stochastic problem (ASP) that

$$l(\gamma) = \mathbf{P}_{\mathbf{u}}(S(\mathbf{w}) \geq \gamma) = \mathbf{E}_{\mathbf{u}} I_{\{S(\mathbf{w}) > \gamma\}} \tag{9.23}$$

which is our estimate of l for γ close to γ^* and typically $S(\mathbf{w}) > \gamma$ is a rare event. We assume the demixing weight, $\mathbf{w}_i$, is sampled from the Gaussian distribution with mean $\mathbf{m}$ and variance β^2. With the CE algorithm, we make adaptive changes to the probability density function, specifically the mean $\mathbf{m}$ and the variance β^2 and the update rule for the parameters is in the form of

$$\hat{\mathbf{m}} = \frac{\sum_{i=1}^{N} I_{\{S(\mathbf{w}_i) \geq \hat{\gamma}_t\}} \mathbf{w}_i}{\sum_{i=1}^{N} I_{\{S(\mathbf{w}_i) \geq \hat{\gamma}_t\}}} \tag{9.24}$$

and

$$\hat{\beta}^2 = \frac{\sum_{i=1}^{N} I_{\{S(\mathbf{w}_i) \geq \hat{\gamma}_t\}} (\mathbf{w}_i - \hat{\mathbf{m}})^2}{\sum_{i=1}^{N} I_{\{S(\mathbf{w}_i) \geq \hat{\gamma}_t\}}}. \tag{9.25}$$

Moreover, we wish to maintain the online update capabilities of cross entropy and so we use the batch-elite method described above rather than an instantaneous gradient ascent update. Actually [140] has already moved some way towards this in that a "smoothed version" of the parameter updating is used:

$$\mathbf{v}_t = \alpha \hat{\mathbf{v}}_t + (1 - \alpha)\hat{\mathbf{v}}_{t-1} \tag{9.26}$$

where α is called the *smoothing parameter* with $0.7 \leq \alpha \leq 1$

Thus we summarize our ICA algorithm with CE method as follows:

1. Initialize $\mathbf{m}_0$ and β_0^2. Set t=1.
2. Generate random samples, $\mathbf{w}_1, ..., \mathbf{w}_N$ from the density $\mathcal{N}(\mathbf{m}_{t-1}, \beta_{t-1}^2)$. From the second independent component onwards, ensure $\mathbf{w}_1, \ldots, \mathbf{w}_N$ is orthogonal to the previous weights, for example simply by using the Gram-Schmidt deflation method.
3. Compute the sample $(1 - \varrho)$ quantile $\hat{\gamma}_t$ of the performance function according to (9.22).
4. Use the "elite" samples and update the reference parameters, $\mathbf{m}$ and β^2 according to (9.24) and (9.25). We denote the solution by $\hat{\mathbf{v}}_t$.
5. Apply (9.26) to smooth out the vector $\mathbf{v}_t$.
6. If not finished, return to step 2.

To demonstrate how well our ICA algorithm works, we use the same 3-D real data set used in Subsection 8.1.1, in which the original signals are pieces of records from different humans speaking as shown in the left column of Figure 9.2. The number of observations is 40000 and we set $\varrho = 0.1, N = 50, \alpha = 0.7$. The number of iterations for each independent component is 20. The recovered independent components are shown in Figure 9.3. The batch learning is performed with the Gram-Schmidt method.

It is clear that all independent components have been separated as shown in Table 9.1 and the Amari error in this case is 0.0115. We find that the ICA algorithm converges so quickly that the simulation time is less than 60 seconds. Note that with the simulation in Subsection 8.1.1, in which even

Fig. 9.2. 3-D real data set used in simulation. Left: the original data set. Right: the mixed observations.

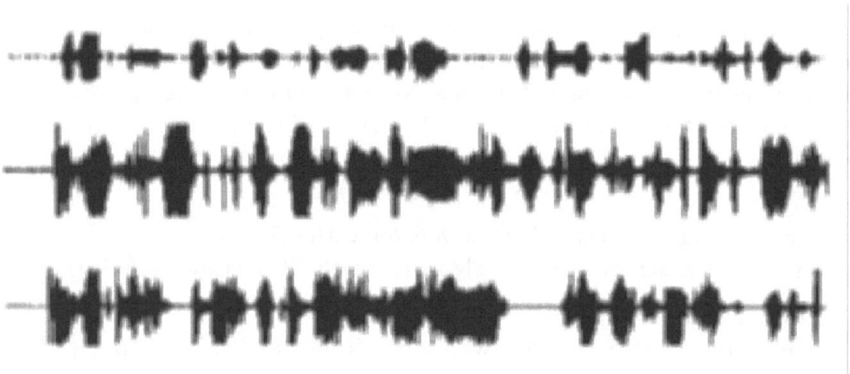

Fig. 9.3. The ICs recovered by the ICA network with 3-D real data set

Table 9.1. The kurtosis of the original signals, mixed observations and recovered independent components (ICs)

	Kurtosis 1	**Kurtosis 2**	**Kurtosis 3**
Original signals	10.7588	9.8449	6.6837
Mixed observations	6.5276	7.2470	8.1544
Recovered ICs	10.7590	9.8428	6.6834

Table 9.2. Correlation between the original sources and recovered signals

1.0000	0.0036	0.0008
0.0045	**1.0000**	0.0094
0.0031	0.0099	**0.9999**

when the number of observations is only 10000, the simulation time was around 15 minutes. Moreover, we can see that the CE algorithm has identified all the independent components with high accuracy from Table 9.2.

9.3 Linear Projection with Cross Entropy Method

In the previous section, we have used the ICA problem as the associated stochastic problem, which can be solved with the cross entropy method. In this section, we wish to investigate the use of the on-line cross entropy method on other projection problems.

The RL machine gets rewards from the external environment depending on how well it has done with an exploratory action and it must choose actions that tend to increase the long-term sum of values of rewards. In the previous section, we use the cross-entropy rationale for the same purpose. There is a clear set of similarities between reinforcement learning and the cross-entropy method:

1. In both, we generate samples or actions from the current estimated distributions but do not have a set of target answers for these samples/actions.
2. In both, we get an estimated reward or performance function value from the generated samples or actions.

So where we previously used a reward function $r = \mathbf{r}(s)$ for reinforcement learning, we may now use the corresponding performance function in the cross entropy method. We use stochastic units drawn from a Gaussian distribution to sample a variety of network weights, which means the weights $\mathbf{W}$ are all drawn from $\mathcal{N}(\mathbf{m}, \beta^2 I)$, the Gaussian distribution with mean $\mathbf{m}$ and variance β^2.

Thus, at iteration t, we create an initial M samples, $\mathbf{w}_1^t, ..., \mathbf{w}_M^t$ from $\mathcal{N}(\mathbf{m}, \beta^2 I)$, a $D-$dimensional isotropic Gaussian distribution with centre, $\mathbf{m}$. We calculate the performance values for each of these samples and put the samples in order of increasing performances, $\mathbf{w}_{p(1)}^t, ..., \mathbf{w}_{p(M)}^t$ where $p(i) \in 1, ..., M$ and select the r highest performing samples, $\mathbf{w}_{p(M-r)}^t, ..., \mathbf{w}_{p(M)}^t$. In this Section, we move the distribution closer to these 'elite' samples, with the learning rules

$$\mathbf{m} \leftarrow (1-\eta)\mathbf{m} + \eta(\frac{1}{r}\sum_{j=M-r}^{M} \mathbf{w}_{p(j)}^t - \mathbf{m}) \tag{9.27}$$

$$\beta^2 \leftarrow (1-\eta)\beta^2 + \eta(\frac{1}{r}\sum_{j=M-r}^{M} (\mathbf{w}_{p(j)}^t - \mathbf{m})(\mathbf{w}_{p(j)}^t - \mathbf{m})^T) \tag{9.28}$$

where η is the learning rate. We then create another M samples, $\mathbf{w}_1^{t+1}, ..., \mathbf{w}_M^{t+1}$ from the new distribution, and iterate the process. We point out that η is the parameter which we must set to ensure the adaptive update of the distribution converges more smoothly.

We summarize our algorithm as follows:

1. Select one item of data from the data set randomly.
2. Generate a sample $\mathbf{w}_{t,1}, ..., \mathbf{w}_{t,M}$ from the currently estimated distribution $\mathcal{N}(\mathbf{m}_t, \beta_t^2)$ and compute the sample $1-\varrho$-quantile $\hat{\gamma}_t$ of the performance according to the performance function.

3. Update the parameters of the distribution to make a new probability density function with (9.28) and (9.27).
4. Go back to 1 until the number of iteration defined.

9.3.1 Principal Component Analysis

To perform PCA, we use the performance function $r = \frac{1}{1+\exp(-\gamma|\mathbf{w}^T\mathbf{x}|)}$. To identify multiple components, in this subsection, we use the Gram-Schmidt method as the deflation method. Thus, from the second component, right after generating a sample $\mathbf{w}_j$ from the currently estimated distribution $\mathcal{N}(\mathbf{m}_t, \beta_t^2 I)$, we subtract $(\mathbf{w}_j^T\mathbf{m}_k)\mathbf{m}_k, k = 1, 2, \ldots, j-1$ from $\mathbf{w}_j$.

To illustrate our algorithm, we create a 5-dimensional artificial data set of 1000 samples, whose elements are drawn independently from Gaussian distributions with $x_i \sim \mathcal{N}(0, i^2)$, so x_5 has the greatest variance and x_1 has the lowest variance. We can see that the five principal components have been identified clearly as shown in Table 9.3 and our algorithm converges smoothly as shown in Figure 9.4.

Table 9.3. The weights from the artificial data experiment for five principal components by the cross entropy method

PC1	0.0144	0.0086	-0.0061	-0.0282	**-0.9994**
PC2	0.0136	-0.0411	-0.0111	**0.9986**	-0.0283
PC3	-0.0127	-0.0076	**0.9998**	0.0108	-0.0067
PC4	0.0418	**0.9982**	0.0077	0.0409	0.0080
PC5	**0.9989**	-0.0414	0.0126	-0.0147	0.0144

Deflationary orthogonalization in performance function

Another way in which to identify multiple components is to integrate deflationary orthogonalization with the performance function directly which is a similar idea to that in Subsection 8.1.4. We re-define the performance function as

$$S(\mathbf{w}) = \begin{cases} \frac{1}{1+exp(-\gamma|\mathbf{w}_1^T\mathbf{x}|)} & \text{if j=1} \\ \frac{1}{1+exp(-\gamma|\mathbf{w}_j^T\mathbf{x}-\Sigma_j^T\mathbf{x}|)} & \text{if j>1} \end{cases} \tag{9.29}$$

where $\Sigma_j = \sum_{k=1}^{j-1} \mathbf{w}_j^T \mathbf{w}_k \mathbf{w}_k$.

We demonstrate deflationary orthogonalization in the performance function with the same data set. We can see that the five principal components have been identified clearly as shown in Table 9.4.

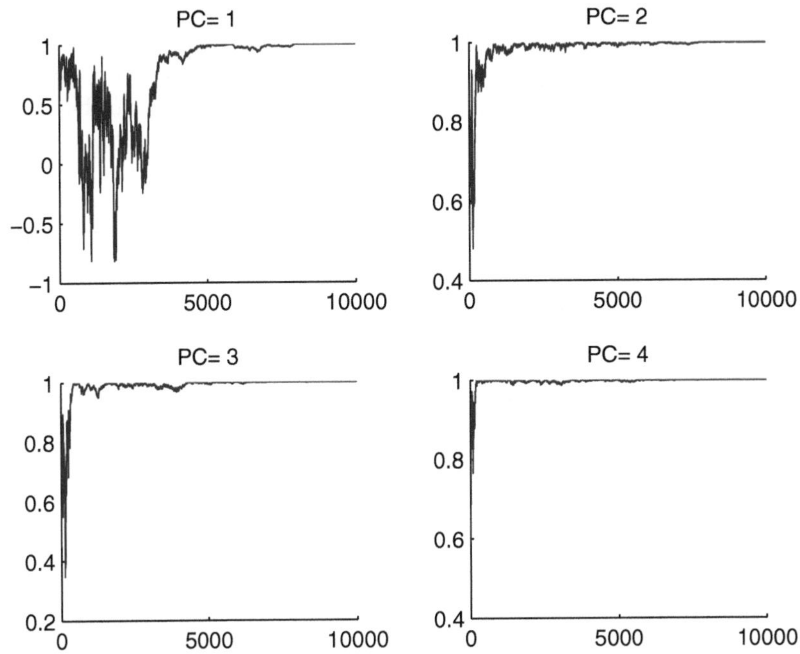

Fig. 9.4. Convergence of the PCA weight vectors to the optimal directions by the on-line CE method

Table 9.4. The weights from the artificial data experiment for five principal components by the cross entropy method. Deflationary orthogonalization is integrated into the performance function.

PC1	-0.0027	0.0052	0.0222	0.0291	**0.9980**
PC2	0.0123	0.0710	0.0200	**0.9937**	0.0480
PC3	0.0608	-0.0531	**-0.9946**	0.0190	0.0376
PC4	0.0533	**0.9936**	-0.0365	-0.1061	-0.0042
PC5	**0.9966**	-0.0697	0.0928	0.0028	0.0130

9.3.2 Exploratory Projection Pursuit

To illustrate our method for exploratory projection pursuit, we create 1000 samples of 5 dimensional data in which 4 elements of each vector are drawn from $\mathcal{N}(0,1)$, while the fifth contains data with negative kurtosis: we draw this also from $\mathcal{N}(0,1)$, but randomly add or subtract 5. Before performing the algorithm, we sphere the data set by eliminating the mean of the data set and dividing by the inverse of the square root of its variance to give a data set that has zero mean and unit variance in all directions. We use $S(\mathbf{w}) = |\tanh(\mathbf{w}^T\mathbf{x})|$ as the performance function. Note that $g(s) = \tanh(s)$ has a expansion of $s - \frac{s^3}{3} + \frac{s^5}{15} \dots$, which is an odd function to measure skewness. However $g(s) = |\tanh(s)|$ is an even function that can be used to measure kurtosis. Table 9.5 shows the outcome of the simulation. We can see that the distribution with negative kurtosis has been identified with high accuracy by the defined performance function. We can also see very stable and quick convergence, as shown in Figure 9.5, to the optimal direction.

Table 9.5. The weights from the artificial data experiment for EPP by the on-line cross entropy method

EC1	-0.0023	0.0028	-0.0019	0.0051	**1.0000**

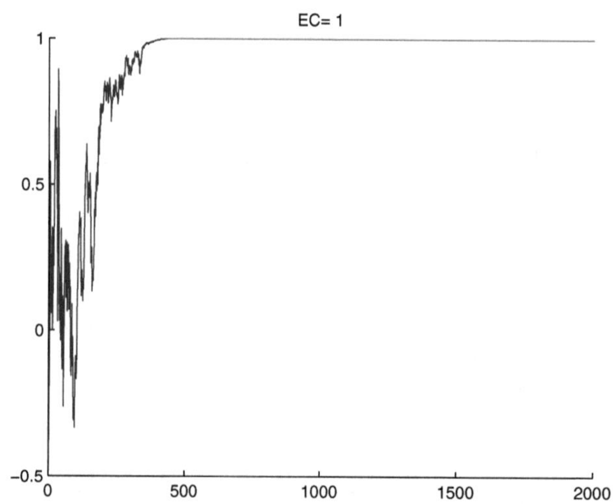

Fig. 9.5. Convergence of EPP simulation by the cross entropy method

9.3.3 Canonical Correlation Analysis

Canonical Correlation Analysis (CCA) is used when we believe there is some underlying relationship between two data sets. Instead of using multi-reward functions as we have done in reinforcement learning, we define the performance function for the first canonical correlation as

$$S(\mathbf{w}) = \frac{1}{1 + \exp(|\mathbf{w}_{1j}^T\mathbf{x}_1 - \mathbf{w}_{2j}^T\mathbf{x}_2| + |(\mathbf{w}_{1j}^T\mathbf{x}_1)^2 - 1| + |(\mathbf{w}_{2j}^T\mathbf{x}_2)^2 - 1|)}, \quad j=1 \tag{9.30}$$

To identify multiple canonical correlations, we eliminate the parts that contain information of the components previously estimated from the current component vector being estimated. Thus for $j > 1$, we have

$$S_0(\mathbf{w}) = \frac{1}{1 + \exp(|\mathbf{w}_{1j}^T\mathbf{x}_1 - \Sigma_{1j}^T\mathbf{x}_1 - \mathbf{w}_{2j}^T\mathbf{x}_2 + \Sigma_{2j}^T\mathbf{x}_2| + |(\mathbf{w}_{1j}^T\mathbf{x}_1)^2 - 1| + |(\mathbf{w}_{2j}^T\mathbf{x}_2)^2 - 1|)}. \tag{9.31}$$

where $\Sigma_{1j} = \sum_{k=1}^{j-1} \mathbf{w}_{1j}^T\mathbf{w}_{1k}\mathbf{w}_{1k}$ and $\Sigma_{2j} = \sum_{k=1}^{j-1} \mathbf{w}_{2j}^T\mathbf{w}_{2k}\mathbf{w}_{2k}$. In practice, we define the performance function in the form of

$$S(\mathbf{w}) = \frac{1}{1 + \exp(|\mathbf{w}_{1j}^T\mathbf{x}_1 - \mathbf{w}_{2j}^T\mathbf{x}_2| + |\Sigma_{1j}^T\mathbf{x}_1 - \Sigma_{2j}^T\mathbf{x}_2| + |(\mathbf{w}_{1j}^T\mathbf{x}_1)^2 - 1| + |(\mathbf{w}_{2j}^T\mathbf{x}_2)^2 - 1|)} \tag{9.32}$$

Thus maximizing (9.32) can guarantee the maximum of (9.31).

To illustrate this method, we use the same artificial data set which was used in Subsection 8.1.3. The number of data is 1000 and the learning rate is 0.025, which is reduced linearly to zero. We illustrate the convergence of the CCA weight vectors in Figure 9.6: we can see consistent convergence towards the first two canonical correlation vectors smoothly. Table 9.6 shows that our algorithm works well in identifying different canonical correlations.

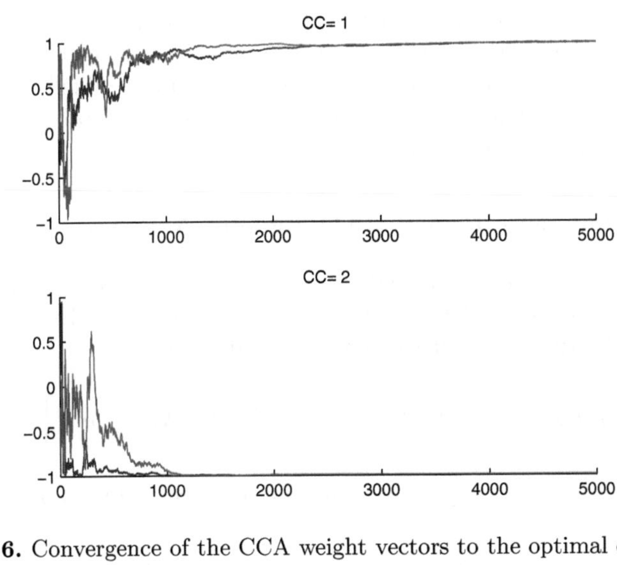

Fig. 9.6. Convergence of the CCA weight vectors to the optimal directions

Table 9.6. The first two Canonical Correlation (CC) weight vectors found for the artificial data

	m1	m2
1st CC	**0.9950**	**0.9905**
	0.0066	-0.1370
	0.0817	0.0113
	-0.0562	
2nd CC	-0.0573	-0.0240
	-0.9914	**-0.9996**
	0.1136	0.0726
	0.0296	

9.4 Cross Entropy Latent Variable Models

Latent Variable Models have been described in Section 3.3. The most common approach for Latent Variable Models is the maximization of the likelihood function using the EM algorithm. In this section we derive a new method for the optimization of the likelihood in latent variable models that is based on the *cross entropy* (CE) method. For latent variable models, we turn the optimization of the likelihood function into the associated stochastic problem (ASP) first. The basic method is

- Generate random samples from the associated stochastic problem using some randomization method.
- Update the parameters (which will typically be parameters of the pdf generating the samples) to make the production of better samples that correspond to the parameters to be optimized in the likelihood function next time. For a Gaussian distribution, this results in updates only to the mean, μ_{ce}, and covariance, Σ_{ce}.

9.4.1 Probabilistic Principal Component Analysis

We have discussed how principal component analysis can be solved based on a latent variable model in Subsection 3.3.2. Consider the $\mathbf{x}$-conditional probability distribution over the data space given by

$$p(\mathbf{t}|\mathbf{x}) = (2\pi\sigma^2)^{-d/2} \exp\left\{ -\frac{1}{2\sigma^2}||\mathbf{t} - \mathbf{W}\mathbf{x} - \mu||^2 \right\} \tag{9.33}$$

and $p(\mathbf{x})$ is assumed as a *prior* distribution over the latent variables, which is defined by

$$p(\mathbf{x}) = (2\pi)^{-q/2} \exp\left\{ -\frac{1}{2}\mathbf{x}^T\mathbf{x} \right\} \tag{9.34}$$

We then obtain the joint distribution of $\mathbf{t}$ and $\mathbf{x}$

$$p(\mathbf{t}_n, \mathbf{x}_n) = (2\pi\sigma^2)^{-d/2} \exp\left\{ -\frac{1}{2\sigma^2}||\mathbf{t}_n - \mathbf{W}\mathbf{x}_n - \mu||^2 \right\}(2\pi)^{-q/2} \exp\left\{ -\frac{1}{2}||\mathbf{x}_n||^2 \right\} \tag{9.35}$$

Thus, the corresponding log-likelihood is given by

$$\begin{aligned} L_c(\theta) &= \textstyle\sum_n [\ln p(\mathbf{t}_n|\mathbf{x}, \mathbf{W}, \sigma^2) + \ln p(\mathbf{x})] \\ &= \textstyle\sum_n \left[-\frac{q+d}{2} \ln(2\pi) - \frac{d}{2} \ln \sigma^2 - \frac{1}{2\sigma^2}(\mathbf{t}_n - \mathbf{W}\mathbf{x} - \mu)^T(\mathbf{t}_n - \mathbf{W}\mathbf{x} - \mu) - \frac{1}{2}\mathbf{x}_n^T\mathbf{x}_n \right] \\ &= -\frac{N(q+d)}{2} \ln(2\pi) - \textstyle\sum_{n=1}^N \Big\{ \frac{d}{2} \ln \sigma^2 + \frac{1}{2\sigma^2}(\mathbf{t}_n - \mu)^T(\mathbf{t}_n - \mu) \\ &\quad -\frac{1}{\sigma^2}\mathbf{x}_n^T\mathbf{W}^T(\mathbf{t}_n - \mu) + \frac{1}{2\sigma^2}\mathtt{tr}(\mathbf{W}^T\mathbf{W}\mathbf{x}\mathbf{x}^T) + \frac{1}{2}\mathtt{tr}(\mathbf{x}\mathbf{x}^T) \Big\} \end{aligned} \tag{9.36}$$

We omit the first term, $-\frac{N(q+d)}{2}\ln(2\pi)$, and take the expectation of L_c with respect to the distribution $p(\mathbf{x}_n|\mathbf{t}_n, \mathbf{W}, \sigma^2)$, and we have

$$\begin{aligned} E\{L_c\} = -\textstyle\sum_{n=1}^N \Big\{ &\frac{d}{2} \ln \sigma^2 + \frac{1}{2\sigma^2}(\mathbf{t}_n - \mu)^T(\mathbf{t}_n - \mu) \\ &-\frac{1}{\sigma^2}E\{\mathbf{x}_n\}^T\mathbf{W}^T(\mathbf{t}_n - \mu) + \frac{1}{2\sigma^2}\mathtt{tr}(\mathbf{W}^\mathbf{T}\mathbf{W}E\{\mathbf{x}\mathbf{x}^T\}) + \frac{1}{2}\mathtt{tr}(E\{\mathbf{x}\mathbf{x}^T\}) \Big\} \end{aligned} \tag{9.37}$$

where $E\{\mathbf{x}\}$ is defined as the mean of the conditional distribution $p(\mathbf{x}_n|\mathbf{t}_n, \mathbf{W}, \sigma^2)$

$$E\{\mathbf{x}\} = \mathbf{M}^{-1}\mathbf{W}^T(\mathbf{t} - \mu) \tag{9.38}$$

and $E\{\mathbf{x}\mathbf{x}^\mathbf{T}\}$ is defined as the variance of the conditional distribution $p(\mathbf{x}_n|\mathbf{t}_n, \mathbf{W}, \sigma^2)$

$$E\{\mathbf{x}\mathbf{x}^T\} = \sigma^2\mathbf{M}^{-1} + E\{\mathbf{x}\}E\{\mathbf{x}\}^T \tag{9.39}$$

where $\mathbf{M} = \mathbf{W}^\mathbf{T}\mathbf{W} + \sigma^2\mathbf{I}$. Thus the parameters to be estimated are $\mu, \mathbf{W}$ and σ^2. The estimation for μ is given by the mean of the data set and σ^2 can be determined by

$$\sigma^2 = \frac{1}{Nd}\sum_{n=1}^{N}\Big\{(\mathbf{t}_n - \mu)^T(\mathbf{t}_n - \mu) - 2E\{\mathbf{x}_n\}^T\mathbf{W}^T(\mathbf{t}_n - \mu) + \mathtt{tr}(\mathbf{W}^T\mathbf{W}E\{\mathbf{x}\mathbf{x}^T\})\Big\}. \tag{9.40}$$

Therefore, the likelihood function of (9.37) is a function of the parameter $\mathbf{W}$. To optimize $\mathbf{W}$, we incoporate the cross entropy method into the optimization of the likelihood function by considering that (9.37) is the performance function $S(\mathbf{W})$ to be maximized over the parameter θ in the cross entropy method. Denoting the maximum by θ^*, we have

$$\gamma^* = \max_{\theta} S(\theta) \tag{9.41}$$

Thus, by defining a family of pdfs $\{f(.;\mathbf{v}), \mathbf{v} \in \nu\}$ on the parameter θ, we follow [140] to associate with (9.17) the following estimation problem

$$l(\gamma) = \mathbf{P}_v(S(\theta) \geq \gamma) = \mathbf{E}_v I_{\{S(\theta) > \gamma\}} \tag{9.42}$$

where θ is the parameters in (9.37). We use stochastic units drawn from a Gaussian distribution to sample θ, which means the set of parameters θ is drawn from $\mathcal{N}(\mathbf{m}, \beta^2 I)$, the Gaussian distribution with mean $\mathbf{m}$ and variance β^2. To estimate l for a certain γ close to γ^*, we make adaptive changes to the probability density function $\mathcal{N}(\mathbf{m}, \beta^2 I)$ according to the Kullback-Leibler cross-entropy. Thus we create a sequence $f(.;\mathbf{v}_0), f(.;\mathbf{v}_1), f(.;\mathbf{v}_2), \ldots$ of pdfs that are optimized in the direction of the optimal density and for the fixed $\hat{\gamma}_t$ and $\hat{\mathbf{v}}_{t-1}$, we derive the $\hat{\gamma}_t$ from the following program

$$\max_{\mathbf{v}} \hat{D}(\mathbf{v}) = \max_{\mathbf{v}} \frac{1}{\mathbf{N}} \sum_{i=1}^{N} I_{\{S(\theta_i) > \hat{\gamma}_t\}} \ln f(\theta_i; \mathbf{v}) \tag{9.43}$$

Since we use a Gaussian distribution as the base learner, the mean and variance of the elite samples are estimated as

$$\mathbf{m} \leftarrow (1-\eta)\mathbf{m} + \eta \frac{1}{r} \sum_{j=M-r}^{M} \mathbf{w}_{p(j)} \tag{9.44}$$

$$\beta^2 \leftarrow (1-\eta)\beta^2 + \eta(\frac{1}{r} \sum_{j=M-r}^{M} (\mathbf{w}_{p(j)} - \mathbf{m})(\mathbf{w}_{p(j)} - \mathbf{m})^T) \tag{9.45}$$

To illustrate our algorithm, we use the same artificial data set as was used in Subsection 8.1.2. The size of the data set is 1000 and we set the number of iteration as 50. We can see that the first principal component has been identified accurately as shown in Table 9.7 and our algorithm converges quickly as shown in Figure 9.7. We also find that the σ^2 in Table 9.7 is close to that calculated by (3.101), $\sigma^2_{\mathbf{ML}} = \frac{1}{D-q}\sum_{j=q+1}^{D} \lambda_j$, which equals the average variance over the lost dimensions.

Futhermore, given that μ is calculated by the mean of the data set, we can optimize $\mathbf{W}$ and σ^2 simultaneously by considering that (9.37) is the performance function $S(\mathbf{W}, \sigma^2)$ to be maximized over the parameters, $\mathbf{W}$ and σ^2. We can see in Table 9.8 that the first principal component has been identified accurately and Figure 9.8 shows that our algorithm converges quickly.

9.4.2 Independent Component Analysis

MacKay [157] has derived a maximum likelihood algorithm in which independent component analysis is performed based on a latent variable model.

Table 9.7. The weights from the artificial data experiment for the 1st principal component by cross entropy method. The parameters of the performance function are $\mathbf{W}$.

PC1					sigma
0.0028	-0.0160	0.0214	-0.0595	**0.9991**	7.2971

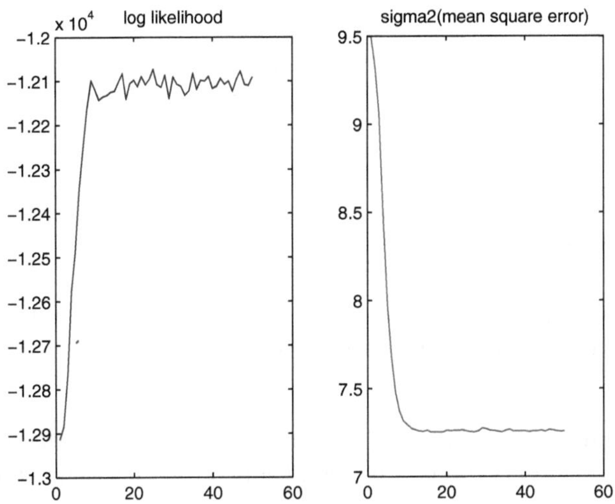

Fig. 9.7. Convergence of the PCA weight vectors to the optimal directions

Table 9.8. The weights from the artificial data experiment for the 1st principal component by the cross entropy method. The parameters in the performance function are $\mathbf{W}$ and σ^2.

PC1					sigma
-0.0007	-0.0081	-0.0369	-0.0830	**0.9958**	6.5497

Defining $\mathbf{V}$ as the mixing matrix and $\mathbf{W}$ as the demixing matrix, the mixed observations $\mathbf{x}$ are generated from latent variables $\mathbf{s}$ that are independently distributed, with marginal distributions $p_i(s_i)$, through the linear mapping $\mathbf{x} = \mathbf{V}\mathbf{s}$. Then the likelihood is given by

$$
\begin{aligned}
P(\mathbf{x}|\mathbf{V}) &= \int P(\mathbf{x}|\mathbf{s})P(\mathbf{s})d\mathbf{s} \\
&= \int \prod_j \delta(\mathbf{x}_j - V_{ij}\mathbf{s}_i) \prod_i p_i(\mathbf{s}_i)d\mathbf{s} \\
&= \frac{1}{\det \mathbf{V}} \prod_i p_i(V_{ij}^{-1}\mathbf{x}_j) \qquad (9.46)
\end{aligned}
$$

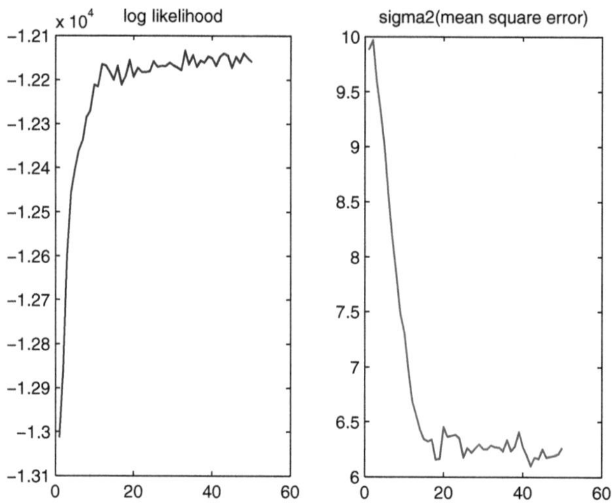

Fig. 9.8. Convergence of the PCA weight vectors to the optimal directions. The parameters in the performance function are $\mathbf{W}$ and σ^2.

where i is the index of the dimension of the sources and j is the index of the dimension of the observations. The log likelihood is then defined as follows

$$\begin{aligned} \log P(\mathbf{x}|\mathbf{V}) &= -\log|\det \mathbf{V}| + \textstyle\sum_i \log p_i(\mathbf{V}_{ij}^{-1} x_j) \\ &= \log|\det \mathbf{W}| + \textstyle\sum_i \log p_i(\mathbf{W}_{ij} x_j) \end{aligned} \tag{9.47}$$

where $\mathbf{W} \equiv \mathbf{V}^{-1}$. [157] has stated that the distribution of the sources can be assumed as $p_i(s_i) \propto \frac{1}{\cosh(s_i)}$. Therefore, we define the performance function as

$$S(\mathbf{W}) = N \log|\det \mathbf{W}| + \sum_{j=1}^{D} \sum_{n=1}^{N} \frac{1}{\cosh(\mathbf{w}_j \mathbf{x}_n)} \tag{9.48}$$

Thus (9.48) is to be maximized over the set of unmixed weight vectors $\mathbf{W}$ that are drawn from $\mathcal{N}(\mathbf{m}, \beta^2 I)$, the Gaussian distribution with mean $\mathbf{m}$ and variance β^2, which is estimated as (9.27) and (9.28). It is worth noting that if we assume that the number of sources is equal to the number of observations, instead of measuring different independent components sequentially by deflation methods, all the independent components can be evaluated simultaneously.

To illustrate our algorithm, we use the 2-dimensional real data set, 'chirp' and 'gong', provided by Matlab and the number of signals is 1000. We can see all the independent components have been identified as shown in Figure 9.9 and our algorithm is highly accurate as shown in Table 9.10. The Amari error in this case is 0.0242.

Table 9.9. The kurtosis of the original signals, mixed observations and recovered independent components (ICs)

	Kurtosis 1	**Kurtosis 2**
Original signals	3.2958	3.9589
Mixed observations	3.8613	3.5871
Recovered ICs	3.2970	3.9547

Table 9.10. Correlation between the original sources and recovered signals

1.0000	0.0276
0.0026	**0.9996**

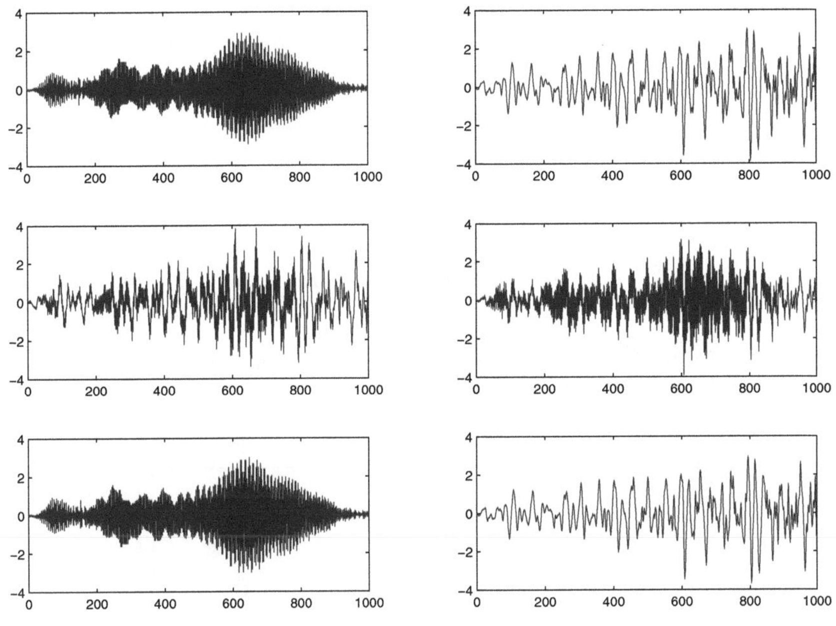

Fig. 9.9. ICA as latent variable model with CE algorithm. Top: the original signals. Middle: mixed observations. Bottom: recovered ICs.

9.4.3 Topology Preserving Manifolds

We follow [37, 96] to create a latent space of points $\mathbf{x}_1, \ldots, \mathbf{x}_K$ which lie equidistantly on a line or at the corners of a grid [233].. To allow non-linear modeling, we map these latent points through a set of M basis functions, typically squared exponentials centered in latent space, and then map the output of the basis functions to points, $\mathbf{m}_1, \mathbf{m}_2, \ldots, \mathbf{m}_K$ through a set of weights. Thus, we have

$$\mathbf{m}_k = \sum_{j=1}^{M} \mathbf{w}_j \Phi_j(\mathbf{x}_k) = \sum_{j=1}^{M} \mathbf{w}_j \exp(-\lambda||\mu_j - \mathbf{x}_k||^2), k = 1, \ldots, K \tag{9.49}$$

where $\phi_j(), j = 1, \ldots, M$ are the M basis functions, and $\mathbf{w}_j$ is the weight from the jth basis function to the data space. The algorithm is:

1. Randomly select a data point, $\mathbf{t}$.
2. Find the closest prototype, say $\mathbf{m}_{k*}$, to $\mathbf{t}$.
3. Generate T samples from the Gaussian distribution, $\mathcal{N}(\mathbf{m}_{k*}, \beta_{k*}^2 I)$. Call the samples, $\mathbf{y}_{k*,1}, ..., \mathbf{y}_{k*,T}$. We note that we are using $\mathbf{m}_1, \mathbf{m}_2, \ldots, \mathbf{m}_K$ to perform two conceptually separate functions, as prototypes or means to which the data will be quantized and as centres of Gaussian distributions from which samples will be drawn.
4. Evaluate the samples using $S(\mathbf{y}) = \exp(-\gamma \parallel \mathbf{y} - \mathbf{t} \parallel^2)$ as the performance function.
5. Sort the samples using $p(1)$ as the worst to $p(T)$ as the best. i.e. we are identifying the r elite samples.
6. Update the parameters

$$\mathbf{w} \leftarrow \mathbf{w} + \eta(\frac{1}{r} \sum_{j=T-r+1}^{T} \mathbf{y}_{k*,p(j)} - \mathbf{m}_{k*})\phi(\mathbf{x}_{k*}) \tag{9.50}$$

$$\beta_{k*}^2 = \beta_{k*}^2 + \eta_0(\frac{1}{r} \sum_{j=T-r+1}^{T} (\mathbf{y}_{k*,p(j)} - \mathbf{m}_{k*})(\mathbf{y}_{k*,p(j)} - \mathbf{m}_{k*})^T) \tag{9.51}$$

   where η, η_0 are the learning rates with typically $\eta = 10\eta_0$.
7. Update the prototypes' positions using (9.49).

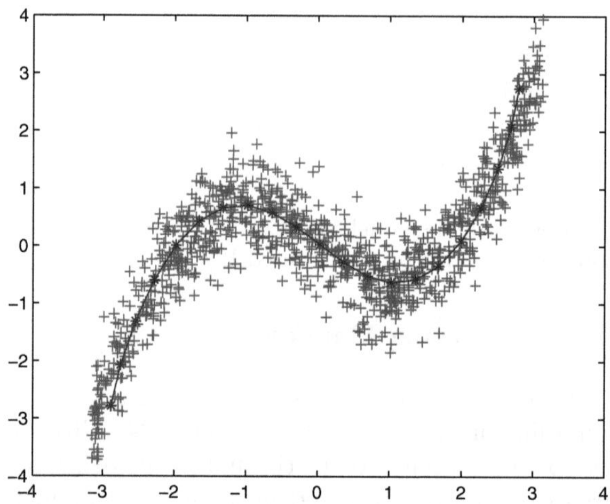

Fig. 9.10. The data are shown by '+'s and the latent points' projections are shown by '*'s

Figure 9.10 shows the result of a simulation in which there are 20 latent points lying equally spaced in a one dimensional latent space, passed through 5 basis functions and mapped to the data space by the linear mapping W. We generate 1000 2-dimensional data points, (x_1, x_2), from the function $x_2 = x_1 + 1.25\sin(x_1) + \rho$, where ρ is the noise from a uniform distribution in $[0, 1]$. The number of iterations is 5000. The latent points' projections are shown in the figure. We clearly see that the one dimensional nature of the data has been identified.

9.5 Deep Architectures in Unsupervised Data Exploration

Deep architectures are compositions of many layers of adaptive non-linear components, which allow representations of wide families of functions in a more compact form than shallow architectures. Bengio and LeCun [27] have demonstrated that deep architectures are often more efficient for solving complicated problems in terms of number of computational components and parameters. A greedy, layer-wise unsupervised learning algorithm [26, 113] has recently been introduced to provide an initial configuration of the parameters with which a gradient-based supervised (backpropagation) learning algorithm is initialized, which results in a very much more efficient learning machine. The idea behind the greedy algorithm is that simpler models are learned sequentially and each model in the sequence recieves a different representation of the data. Thus features produced by the lower layers represent lower-level abstractions, which are combined to form high-level features at the next layer, representing higher-level abstractions. In this section, we wish to investigate deep architectures in unsupervised data exploration.

9.5.1 Multilayer Topology Preserving Manifolds

We extend the topology preserving mapping with cross entropy, described in Subsection 9.4.3, to a multilayer topology preserving mapping. In each layer, we create a q-dimensional latent space with a regular array of points, $\mathbf{X} = (\mathbf{x}_1, \ldots, \mathbf{x}_K)$ that have the structure of lying equidistantly on a line or on a grid. These latent points are nonlinearly mapped to points, $(\mathbf{m}_1, \ldots, \mathbf{m}_K)$ in the input space through a set of basis functions, which forms a set of reference vectors, $\mathbf{m}_i = \mathbf{W}\phi(\mathbf{x}_i)$. Each of the reference vectors then forms the centre of the Gaussian distribution in the input space and we can represent the distribution of the data points in input space in terms of a smaller q-dimensional nonlinear manifold by maximizing the log-likelihood function. Thus we have a higher-level representation of the data points in the input space by the projection of the data points in the latent space. Denoting the latent space representation of each data point as $\mathbf{t}_n^{latent}$, we have

$$\mathbf{t}_n^{latent} = \sum_{i=1}^{K} r_{ni}\mathbf{x}_i \quad (9.52)$$

$$r_{ni} = \frac{\exp(-\gamma d_{ni}^2)}{\sum_k \exp(-\gamma d_{nk}^2)} \quad (9.53)$$

where r_{ni} is responsibility of the ith latent point for the nth data point and $d_{pq} = ||\mathbf{t}_p - \mathbf{m}_q||$, is the euclidean distance between the pth data point and the projection of the qth latent point. The projection of the data points in the latent space of this layer then becomes the data points in the input space of the next layer. Therefore, the topology preserving mapping in each layer performs a non-linear transformation on its input vectors and produces as output the vectors that will be used as input for the topology preserving mapping in the next layer and the projection of the data points in higher layers may represent more abstract features, whereas lower layers extract low-level features from the data set.

To demonstrate our multilayer topology preserving mapping algorithm, we create a deep architecture model with four layers, in each layer of which we use a 2 dimensional grid of latent points: we use a 21×21 grid of latent points being passed through a 5×5 set of basis vectors. We begin by illustrating the algorithm on the well-known wine data set from the UCI Repository of machine learning databases. It has 178 samples, 13 features and 3 classes. Because some of the features are scaled up to 1500 and others lie between 0 and 1, we preprocessed the data by normalizing all features between -1 and 1. We develop a new way to represent the experimental results, which makes it easier for us to evaluate how well the algorithm works: we calculate the sum of the responsibility vectors over the data points, $R = \sum_{n=1}^{N} \mathbf{r}_n$, and then form a responsibility grid using R for each latent point. Since the latent points are fixed in (9.52), the data points belonging to the same cluster will have similar responsibility vectors, and thus the area where one cluster is located in the responsibility grid will become 'hot', which allows us to identify the clusters easily.

A plot of the responsibility grid and the projection of the data points in the latent space in each layer is shown in Figure 9.11. We can see that although the data points have been mapped into the latent space accurately, there are only several small hot areas in the responsibility grid, which means we can not identify different clusters directly. We also find that the higher-level the layer is, the hotter the areas corresponding to different cluters is and in the fourth layer, we can see that the three clusters have been identified clearly and the data points belonging to the same cluster is much closer than in the first layer in the latent space. Therefore, we consider the multilayer topology preserving mapping model has extracted more abstract features in the higher layer. In addition, we can also obviously see that in the fourth layer, the 'square' cluster and the 'cycle' cluster are closer to each other, which are far away from the 'cross' cluster.

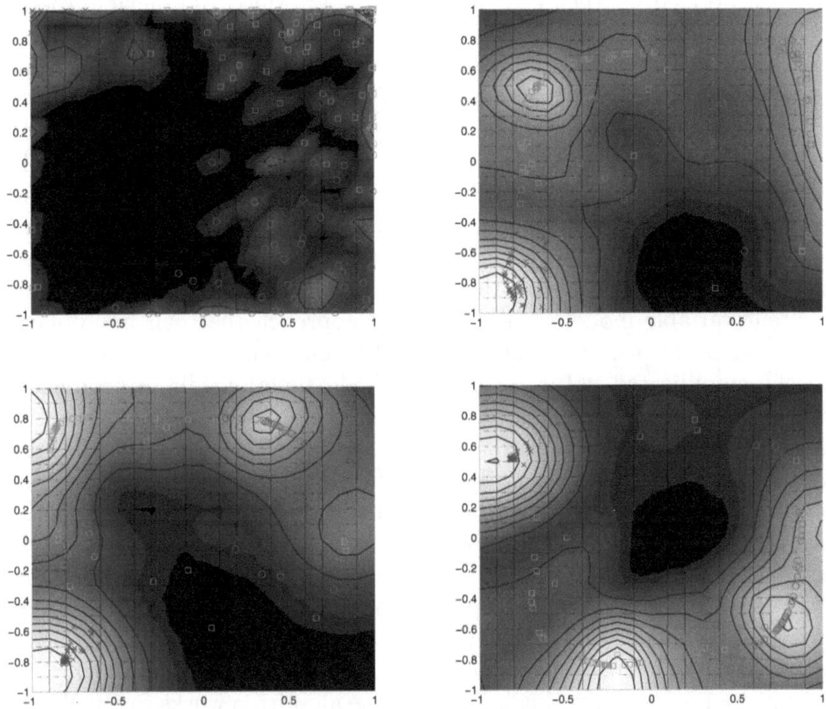

Fig. 9.11. Plot of the responsibility grid and the projection of the data points in the latent space in each layer. Each contour line represents a constant value of R. The cross, square and cycle correspond to the data points for different clusters. Top left: the 1st layer. Top right: the 2nd layer. Bottom left: the 3rd layer. Bottom right: the 4th layer.

This is an illustration of using a novel method of adaptation to a current area of research investigation, deep architectures. We shall have more to say on this matter in the conclusion of the book.

9.6 Conclusion

In this chapter, we have presented a way to apply the cross-entropy method to projection problems. We achieved this goal by regarding such problems as optimization problems that can be turned into rare-event estimation problems and then a variety of appropriate performance functions have been used to solve different projection problems. We have used stochastic units drawn from a Gaussian distribution to sample the component weight vectors. With the CE algorithm, we make adaptive changes to the probability density function of component weight vectors according to the Kullback-Leibler cross-entropy. We have first applied the batch cross entropy method to the ICA problem and then derived another way in which we performed the linear projection methods with an on-line cross entropy method. With the on-line adaptive update rule, the convergence of the probability density function of component weight vectors has become smoother and faster.

We can not say whether the batch or the on-line ways to perform the cross entropy method is better. Although projection problems can be solved in batch or on-line learning, some projection problems are more suitable for performing with batch learning, which can achieve faster convergence with greater accuracy. However, other projection methods have best performance only if the convergence is smooth enough, which comes from on-line learning.

Latent variable modeling is a powerful approach that can be applied to projection problems. However, due to the constraints on the parameters, and the complicated nature of the likelihood function, direct optimization of the likelihood function is not a simple task. Moreover, traditional methods such as gradient-based algorithms or the EM algorithm often requires much human effort in deriving the update rule for the parameters to archive global optimum. We have shown that three latent variable models with cross entropy have converged with high accuracy even without specifying the particular learning rules for the parameters.

We have demonstrated a deep architecture in unsupervised data exploration. We developed multilayer topology preserving manifolds with the cross entropy method in clustering. The results have been expressed in a particular way which allows us to evaluate how well the algorithm works easily. We have shown that the deep architecture algorithm has identified the clusters more accurately.

10

Artificial Immune Systems

10.1 Artificial Immune Systems

The Artificial Immune System paradigm (AIS) is inspired by the biological immune system whose main goal is to protect the human body from the attack of foreign pathogens such as virus, fungus or other parasites. The biological immune system is capable of distinguishing between the normal components of our organism and the foreign materials that can cause us harm that are known as *antigens*. The molecules called antibodies play the main role in the immune system response in that the immune response is specific to a certain antigen. Thus when an antigen is detected, those antibodies that best recognize an antigen will proliferate by cloning. Then the new cloned cells undergo a mutation or hypermutation process so that their receptor population will be increased. These mutations experienced by the clones are inversely proportional to their affinity to the antigen, which means that those antibodies with the highest affinity suffer the lowest mutation rates, whereas the lowest affinity antibodies have high mutation rates. After this mutation process ends, the antibodies' affinity in the immune system is improved and the immune system returns to its normal condition by eliminating the extra cells. However, some cells are turned into memory cells so that when the immune system is later attacked by the same type of antigen (or a similar one), these memory cells are activated, presenting a better and more efficient response. Artificial immune systems are motivated from such immunology in order to develop systems capable of performing a wide range of tasks. They consist of, at least:

- *Antibody*: A real antibody is the portion of a surface receptor molecule on a leukocyte that binds with a specific molecular pattern. B-cells and T-cells are two types of leukocyte. In an artificial immune system, no distinction is usually made between a B-cell and its receptor, hence every element of the artificial immune system is known as an antibody.

W.A. Barbakh et al.: Non-Stand. Param. Adapt. for Explor. Data Analys., SCI 249, pp. 175–197.
springerlink.com

- *Antigen*: The molecular patterns that can be recognized by the surface receptors. In an artificial immune system, antigens are usually represented as a field within a data item with a particular value.
- *Affinity*: The degree of the strength of binding between an antibody and an antigen, which is equivalent to the concept of *fitness* in an evolutionary algorithm.

Castro and Timmis [61] have created abstract models of immune cells and molecules based on the shape-space approach. In addition, a set of general AIS algorithms have been derived, which include a negative selection algorithm [83], a positive selection algorithm [200], a clonal selection algorithm [65], continuous immune network models [60], and discrete immune network models [63, 216]. Also, the immune-inspired algorithms have been applied to a wide range of applications, including clustering [62, 215], pattern recognition [58, 47, 105] and optimization problems [217, 50, 46].

In the following subsection, we give details of the Clonal Selection Algorithm (CLONALG) and Artificial Immune Network (aiNet-opt).

10.1.1 Clonal Selection Algorithm

The clonal selection and affinity maturation principles [3, 45, 44] are at the core of the immune system when reacting to pathogens, which aims to improve the capability of recognizing and eliminating pathogens. In a simple form of clonal selection as shown in Figure 10.1, when a pathogen invades the organism, the portion of immune cells that recognize these pathogens are selected to proliferate, during which these cloned cells suffer affinity mutations. Thus a new population of higher affinity cells in relation to the invading pathogen differentiates into memory cells while a portion of the clones would be eliminated.

Inspired by the clonal selection principle and the affinity maturation process, de Castro [65] has developed a clonal selection algorithm, named CLONALG, to perform pattern recognition and optimization. The main immune aspects in the CLONALG algorithm [65] include:

- maintenance of a specific memory set;
- selection and cloning of the most stimulated antibodies;
- death of non-stimulated antibodies;
- affinity maturation and re-selection of the clones proportionally to their antigenic affinity;
- generation and maintenance of diversity;

For optimization problems, there is no explicit antigen population defined. Instead the antigen is represented by the objective function to be optimized. Also, an antibody affinity corresponds to the evaluation of the objective function for a given antibody. Some notations are shown below:

- **Ab**: antibody repertoire in a population, $\mathbf{Ab} \in \mathcal{S}^{N \times L}$, where N is the size of the population and L is the dimension of the antibody;

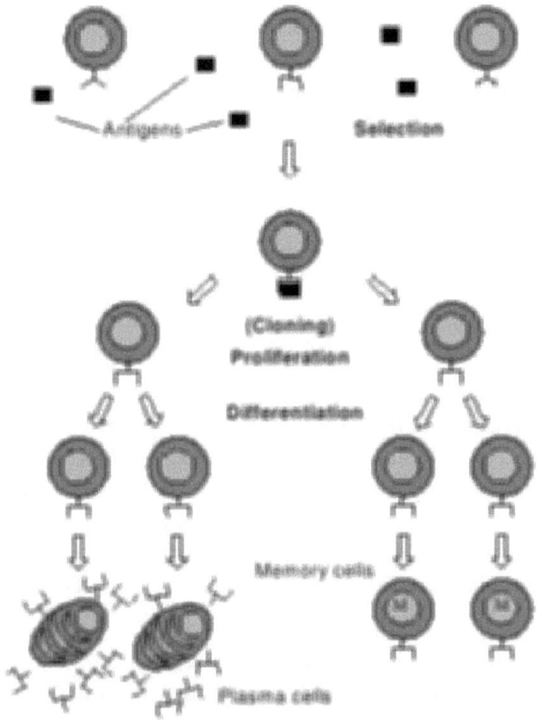

Fig. 10.1. The clonal selection and affinity maturation principles [65]

- **Ag**: population of antigens to be recognized. For optimization, **Ag** is defined to be the objective function to be optimized;
- $\mathbf{f}_i$: the affinity vector of all antibodies in relation to antigen **Ag**;
- $\mathbf{Ab}_{\{n\}}$: n antibodies from the population of antibodies with highest affinities in relation to antigen **Ag**, $\mathbf{Ab}_{\{n\}} \in \mathcal{S}^{n\times L}$;
- **C**: population of N_c clones generated from $\mathbf{Ab}_{\{n\}}$, $\mathbf{C} \in \mathcal{S}^{N_c\times L}$;
- N_c: the size of population of clones **C**;
- $\mathbf{C}^*$: population from **C** after the affinity maturation process;
- $\mathbf{Ab}_{\{d\}}$: set of d new antibodies that will replace d low-affinity antibodies in **Ab**.

The CLONALG algorithm can then be described as follows [65]:

1. Randomly generate a population of antibodies **Ab**.
2. Determine the affinities of all the N antibodies to form the affinity vector $\mathbf{f}_i$.
3. Select the n antibodies with highest affinity from **Ab** to compose a new set $\mathbf{Ab}_{\{n\}}$.

4. Clone the n selected antibodies independently and proportionally to their antigenic affinities to form a repertoire $\mathbf{C}$. The higher the antigenic affinity, the higher the number of clones generated for each of the n selected antibodies. The size of population $\mathbf{C}$ is N_c.
5. The repertoire $\mathbf{C}$ is submitted to an affinity maturation process inversely proportional to the antigenic affinity, generating a population $\mathbf{C}^*$ of matured clones: the higher the affinity, the smaller the mutation rate.
6. Determine the affinity $\mathbf{f}^*$ of the matured clones $\mathbf{C}^*$.
7. From this population of mature clones $\mathbf{C}^*$, select n clones with highest affinities to be the antibodies in the new population $\mathbf{Ab}$ in the next iteration.
8. Finally, replace the $d\%$ lowest affinity antibodies from $\mathbf{Ab}$ by new individuals.

Thus the CLONALG algorithm is capable of learning a set of input patterns by selecting, reproducing and mutating a set of "artificial immune cells". Several affinity maturation ways have been derived in [54, 136]. The convergence analysis of the CLONALG algorithm also has been studied in [224, 55]. [65] has pointed out that there are three user-defined parameters to control the convergence speed and the computational complexity.

- n, the number of antibodies to be selected for cloning. Although n does not strongly influence the number of iterations, it has a strong influence on the size of the population $\mathbf{Ab}_{\{n\}}$. Thus a large value of n leads to a higher computational cost.
- N_c, the size of the population of clones. N_c is defined by $N_c = \sum_{i=1}^{n} round(\beta_i \cdot N)$, where $round(\cdot)$ is the operator that rounds its argument towards the closest integer. The β_i is usually set at 0.1 and the higher N_c, the faster the convergence.
- $d\%$, the rate of low affinity antibodies to be replaced. For values $d > 0$, the algorithm can locate all the peaks of the function to be optimized, and thus high values will lead to a large random search in the affinity space.

10.1.2 Artificial Immune Network

An extension of CLONALG algorithm [63] has been proposed by combining CLONALG algorithm with the immune network theory [126], named aiNet. This immune network was first applied to clustering applications [64] and then an optimization version of aiNet, named *opt-aiNet* was derived in [60] based on the same motivations. In this subsection, the description of artificial immune network is mainly based on the work in [60].

The algorithm of opt-aiNet is summarized as follows:

1. Randomly initialize a population of cells, $\mathbf{Ab}_t$.
2. Determine the fitness vector $\mathbf{f}_i$ of each network cell and normalize the vector of fitnesses.

3. Generate a number N_c of clones for each network cell.
4. Mutate each clone inversely proportional to the fitness of its parent cell to form the population of clones. The parent cells are kept in the population. The maturation is defined as,

$$
\begin{aligned}
c_i^{'} &= c_i + \alpha \mathcal{N}(0,1) \\
\alpha &= \frac{1}{\beta} \exp(-\mathbf{f}_i)
\end{aligned}
\tag{10.1}
$$

5. Determine the fitness of all individuals of the population.
6. For each clone, select the cell with highest fitness and calculate the average fitness of the selected population.
7. If the average fitness of the population is not significantly different from the previous iteration, then continue. Else, return to step 2.
8. Determine the affinity of all cells in the network. Suppress all but those cells with the highest fitness, i.e. those whose affinities are less than the suppression threshold σ_s and determine the number of network cells, named memory cells, after suppression.
9. Introduce a percentage $d\%$ of randomly generated cells.
10. If the variation of the number of cells from one suppression to another is less than the threshold, then the stopping criterion is met, otherwise go back to step 2.

[60] noted that a significant difference between this opt-aiNet and the CLONALG algorithm is that the algorithm described above not only utilizes steps in the CLONALG algorithm, but also involves an interaction among the cells in the immune network and between the network cells and the environment. The aiNet algorithm firstly optimizes the network cells through the clonal selection and affinity maturation process, then when the population reaches a stable state, interaction among the network cells is activated to eliminate some of the similar cells to avoid redundancy. Furthermore, [59] has pointed out that metadynamics is a remarkable feature of the artificial immune network. Besides the maturation of the immune response defined by (10.1), the structure plasticity of the network is defined as

$$
\Delta \mathbf{Ab} = N_c - N_s + N_b - N_d \tag{10.2}
$$

where N_c corresponds to the size of clone population; N_s is denoted as the number of antibodies suppressed due to the interactions; N_b represents the number of incoming antibodies to the network; N_d corresponds to the number of antibodies removed from the network. It is also worth noting that new cells would come into the population of cells only if the current cells can not improve the average fitness greatly. Another difference is that unlike the affinity maturation with a control strategy in the CLONALG algorithm, a Gaussian maturation is introduced which is inversely proportional to the normalized fitness of each parent cell.

However, Timmis and Edmonds [217] have also pointed out that considering the suppression process where for two cells whose Euclidean distance is less than the suppression threshold, the first of which will be suppressed, which may result in deletion of a potential optimum solution. To avoid the chance of this occuring, it [217] has been suggested that a simple ordering can be placed on the network cells and those with least fitness would be removed.

In this chapter, we develop immune-inspired algorithms based on artificial immune systems. Given the data set represented by antigens, all the algorithms perform the same goal: the antibodies are generated in order to improve their affinity defined in different forms according to different projection problems. We begin by extending the well-known CLONALG algorithm [3] to projection problems, in which a little modification is made to the way that we perform the affinity proportional maturation in an artificial immune network. Then we continue to extend the modified CLONALG algorithm in which multiple components can be identified directly in the immune system itself.

We then continue to improve the performance of the non-standard adaptation methods by combining the algorithms we have derived or extended in this book. We point out that the "better performance" is not limited to improving the accuracy of the final results, but also reducing the necessary size of data set and the number of iterations required to achieve the global optimum. We will demonstrate how well the combined algorithms work in comparision to the results of their parent algorithms. Two combined algorithms are developed. One is to incorporate cross entropy into the immune system and the other is to integrate the immune-inspired algorithm with the Q-learning method.

Finally, we combine the three non-standard adaptation methods by bagging. A new way is presented to determine the final solution based on the quality of the local solutions from the bags. The experimental results show that this combined method converges to the optimal solution more reliably and stably and with higher accuracy than the individual methods.

10.2 Projection with Immune-Inspired Algorithms

In this section, we extend artificial immune systems to solving projection problems. We demonstrate that we can apply the CLONALG algorithm, described in Subsection 10.1.1, to projection problems with a little modification. Keeping in mind that it is usually expected to identify multiple components, we present a new way which incorporates deflationary orthogonalization into the immune algorithm directly. We show that our method is quite general and can be easily applied to different projection methods.

10.2.1 Linear Projections with the Modified CLONALG Algorithm

We first consider the immune-inspired algorithm for the problem of independent component analysis. In an artificial immune system, we define the mixed observations to be the antigens, $\mathbf{Ag}$. Each antibody Ab_i in a population of antibodies $\mathbf{Ab}$ represents a possible solution of the k-th independent component filter $\mathbf{w}_k$. One of the most common principles for ICA is to make all independent components as non-Gaussian as possible. Thus an antibody affinity corresponds to the evaluation of the absolute value of the kurtosis of the distribution of the recovered signals, $\mathbf{Wx}$, where $\mathbf{x}_i$ corresponds to an antigen Ag_i.

We consider a batch-learning CLONALG algorithm. Thus given the antigens, those antibodies with high affinities will be selected and cloned. Then these clones suffer an affinity maturation process, in which those antibodies with the highest affinity suffer the lowest mutation rates, whereas the lowest affinity antibodies have high mutation rates. After this mutation process, those matured clones with high affinities are kept in the new population of antibodies that will be used in the next iteration. Therefore, with the antibodies' affinity in the immune system improved, the demixing matrix will be optimized. We define the affinity as

$$\mathbf{f}_i = \left|\mathtt{kurt}(Ab_i^T \mathbf{X})\right|. \tag{10.3}$$

The modified CLONALG algorithm in detail is summarized as follows:

1. Initialize each antigen so that it corresponds to each observation in the data set. The size of the data set is N.
2. Randomly generate a population of real-value antibodies $\mathbf{Ab}^{(t)}$. The size of the population is $N_{\mathbf{Ab}}$.
3. Determine the affinities to all the $N_{\mathbf{Ab}}$ antibodies to form the affinity vector $\mathbf{f}$.
4. Select the n antibodies with highest affinity from $\mathbf{Ab}^{(t)}$ to compose a new set $\mathbf{Ab}_{\{n\}}$.
5. Clone the n selected antibodies independently to form a repertoire $\mathbf{C}$. The number of clones for each antibody in $\mathbf{Ab}_{\{n\}}$ is $round(\beta \cdot N)$, thus the size of population of clones is $N_c = \sum_{i=1}^{n} round(\beta \cdot N) = n \cdot round(\beta \cdot N)$.
6. The population of clones $\mathbf{C}$ is submitted to an affinity maturation process. Each clone is matured by $c_{ij}^{'} = c_{ij} + \alpha \exp(-\mathbf{f}_i)\mu$, where μ is drawn from a Gaussian distribution with zero mean and unit variance, $\mu \sim \mathcal{N}(0,1)$ and c_{ij} is defined as the jth clone of the ith parent antibody. We have a new population $\mathbf{C}^*$ of matured clones.
7. Determine the affinity $\mathbf{f}^*$ of the matured clones $\mathbf{C}^*$.
8. From this population of mature clones $\mathbf{C}^*$, select n clones with the highest affinities to be the antibodies in the new population $\mathbf{Ab}^{(t+1)}$ in the next iteration.
9. Finally, replace the $(1 - \frac{n}{N_{\mathbf{Ab}}}) \times 100\%$ lowest affinity antibodies in $\mathbf{Ab}^{(t)}$ by the new individuals.

There are a few modifications from the classic CLONALG algorithm described in Subsection 10.1.1. While the classic CLONALG algorithm encodes the individuals of the population using binary strings, the algorithm here is

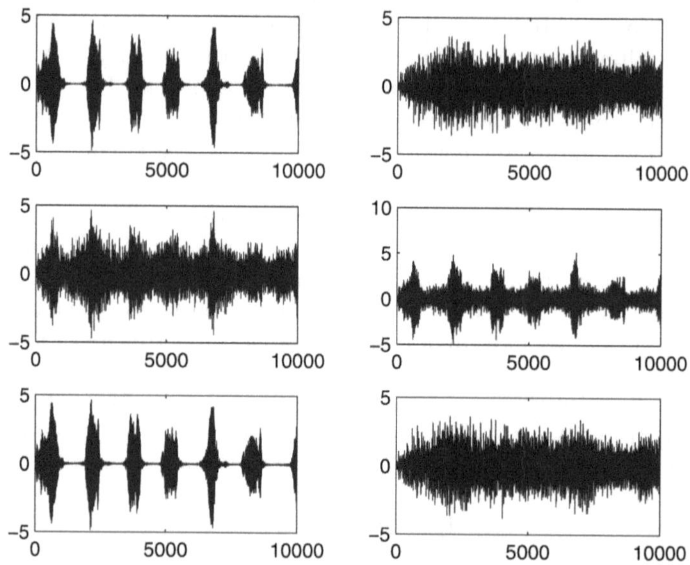

Fig. 10.2. ICA with modified CLONALG algorithm. Top: the original signals. Middle: mixed observations. Bottom: recovered ICs.

based on real-valued vectors. We also apply affinity proportional maturation described in Subsection 10.1.2. In our case, after the n clones with highest affinities are selected for the new population $\mathbf{Ab}^{(t+1)}$, we generate another $(N_{\mathbf{Ab}} - n)$ new individuals. We therefore optimize the demixing matrix $\mathbf{W}$ by adjusting the position of the antibodies in the search space to reach the global optimum.

To illustrate the modified CLONALG algorithm for ICA, we use the 2-dimensional real data set, 'chirp' and 'gong', provided by Matlab. The number of original signals is 10000 and we set $N_{\mathbf{Ab}} = 100, n = 10, \beta = 0.1$. So $N_c = 100$. The number of iterations for the estimation of each independent component is 10. The Gram-Schmidt method is performed to identify all the independent components. In Figure 10.2, we see that all the independent components have been identified with extremely high accuracy as shown in Table 10.1 and Table 10.2, in which the Amari error is only 0.0035. Moreover, Figure 10.3 shows that although the number of iterations is 10, the modified CLONALG algorithm can identify the independent component as early as the first two iterations, and so the speed of convergence is also extremely fast.

Therefore, in the modified CLONALG algorithm, the clones belonging to the same parent antibody Ab_i are samples from the Gaussian distribution with mean Ab_i and variance $\exp(-\mathbf{f}_i)$. We also find from the experiments that a smaller variance when cloning and maturing the parent antibody leads

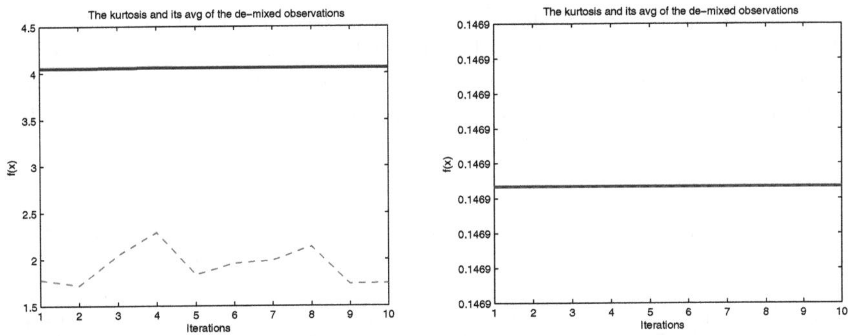

Fig. 10.3. Affinity of the population. Highest (solid line) and average (dashed line). Left: affinity for the 1st IC. Right: affinity for the 2nd IC.

Table 10.1. The kurtosis of the original signals, mixed observations and recovered independent components (ICs)

	Kurtosis 1	**Kurtosis 2**
Original signals	7.0592	3.1467
Mixed observations	3.4269	5.2805
Recovered ICs	7.0593	3.1469

Table 10.2. Correlation between the original sources and recovered signals

	Signal 1	Signal 2
Recovered signal 1	**1.0000**	0.0014
Recovered signal 2	0.0003	**1.0000**

to higher accuracy and smoother convergence. Thus, in practice, the parent antibodies are cloned and matured by

$$c'_{ij} \sim \mathcal{N}(Ab_i, \alpha \exp(-\mathbf{f}_i)), \alpha \in (0, 1]. \tag{10.4}$$

We apply the modified CLONALG algorithm to solve principal component analysis. We use the same 5 dimensional data as in Subsection 8.1.2 in which the first principal component is readily identified as the fifth input dimension. The goal of PCA is to find the linear projection of a data set which contains maximal variance, thus the affinity between the antigens and the antibody is defined as

$$\mathbf{f}_i = \frac{1}{1 + \exp(-\gamma(\mathbf{Ab}_i^T \mathbf{X})^2)}. \tag{10.5}$$

The size of the data set is 10000 and we set $N_{\mathbf{Ab}} = 100, n = 10, \beta = 0.1, \alpha = 0.5, \gamma = 0.00001$. The number of iterations is 100. Table 10.3 shows that the modified CLONALG algorithm has identified the first principal component with high accuracy and we see in Figure 10.4 that the algorithm can identify the first principal component within five iterations and then climbs towards the global optimum. In addition, we can see in Table 10.3 and Table 10.4 that the accuracy is improved by setting a smaller variance in (10.4).

Table 10.3. The first principal component with 5-dimensional artificial data by CLONALG algorithm with $\alpha = 0.7$

PC1	-0.0168	0.0153	0.0122	0.0369	**0.9992**

Table 10.4. The first principal component with 5-dimensional artificial data by the modified CLONALG algorithm with $\alpha = 1$

PC1	-0.1030	0.1022	-0.1328	-0.0213	**0.9802**

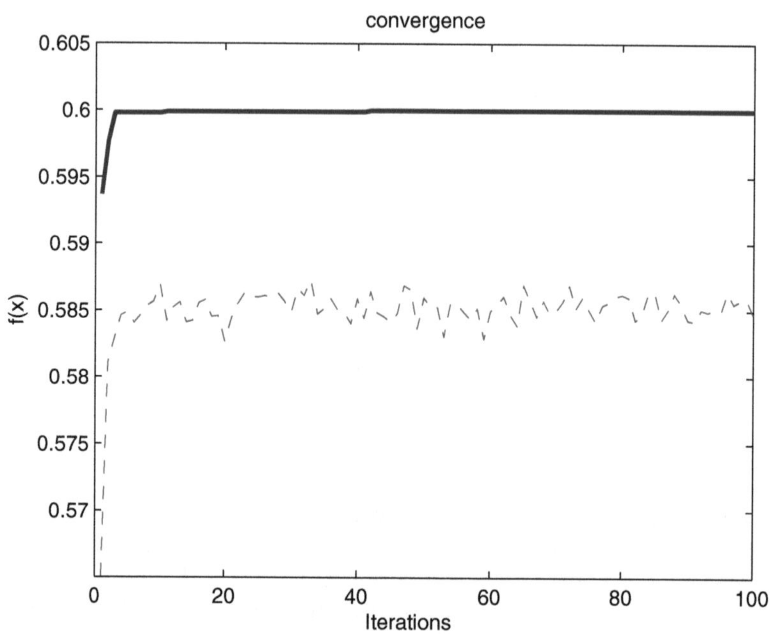

Fig. 10.4. Affinity of the population by the modified CLONALG algorithm for PCA. The vertical axis shows the affinity in each iteration. The horizontal axis shows the number of iterations. Solid line: the highest affinity. Dashed line: the averaged affinity.

10.2.2 Multiple Components

It is straightforward to apply the Gram-Schmidt method to identify multiple components. Thus assuming the first $j-1$ weight vectors have been estimated, we modify step 2 in the modified CLONALG algorithm, where for each antibody Ab_i^j, we perform $Ab_i^j \leftarrow Ab_i^j - \sum_{k=1}^{j-1}(Ab_i^j \mathbf{w}_k)\mathbf{w}_k$, where $\mathbf{w}_k$ corresponds to the kth weight vector previously estimated and Ab_i^j corresponds to the ith antibody for the current weight vector. Then by using the same data set in the previous subsection, Table 10.5 shows that all the principal components have been identified.

Table 10.5. The weights from the artificial data experiment for five principal components with the Gram-Schmidt method

PC1	-0.0168	0.0153	0.0122	0.0369	**0.9992**
PC2	-0.0211	-0.0276	0.1082	**0.9927**	-0.0379
PC3	-0.0225	0.0225	**0.9929**	-0.1084	-0.0088
PC4	-0.0187	**0.9986**	-0.0201	0.0289	-0.0164
PC5	**0.9968**	0.0188	0.0245	0.0197	0.0154

However, we are more interested in identifying multiple components within the artificial immune system itself. We consider each weight vector $\mathbf{w}_i$ previously estimated as a memory cell in a population $\mathbf{Ab}_{\{m\}}$ in which $Ab_{\{m\}}^k$ corresponds to the kth memory cell in the population. The distance between one antibody Ab_i and these memory cells in population $\mathbf{Ab}_{\{m\}}$ can be measured, so that a low distance implies that the antibody Ab_i is similar to the weight vectors previouly estimated. Thus the basic idea is that the antibodies selected to be cloned and matured should have high distances from the memory cells.

We re-structure the modified CLONALG algorithm for multiple components: at iteration t, we first evaluate the distance between each antibody in $\mathbf{Ab}$ generated in step 2 and the memory cells in $\mathbf{Ab}_{\{m\}}$. The $N'_{\mathbf{Ab}}$ antibodies with highest distance are selected and others are eliminated from the population. Then n antibodies with highest affinity from the $N'_{\mathbf{Ab}}$ antibodies are re-selected to be cloned and matured. The modified CLONALG algorithm for the kth component is summarized as follows:

1. Initialize each antigen corresponding to each data point in the data set. The size of the data set is N.
2. Randomly generate a population of real-valued antibodies $\mathbf{Ab}^{(t)}$. The size of the population is $N_{\mathbf{Ab}}$.
3. Measure the distances between each antibody Ab_i and the memory cells $\mathbf{Ab}_{\{m\}}$ according to (10.6) and select the $N'_{\mathbf{Ab}}$ antibodies with lowest distance. For the first component, this step is omitted and $N'_{\mathbf{Ab}} = N_{\mathbf{Ab}}$.

4. Determine the affinities to all the $N'_{\mathbf{Ab}}$ antibodies to form the affinity vector $\mathbf{f}_i$.
5. Select the n antibodies with highest affinity to compose a new set $\mathbf{Ab}_{\{n\}}$.
6. Mature and clone the n selected antibodies independently according to (10.4) to form a population $\mathbf{C}^*$. The size of the population of clones is $N_c = \sum_{i=1}^{n} round(\beta \cdot N'_{\mathbf{Ab}}) = n \cdot round(\beta \cdot N'_{\mathbf{Ab}})$.
7. Determine the affinity $\mathbf{f}^*$ of the matured clones $\mathbf{C}^*$.
8. From this population of mature clones $\mathbf{C}^*$, select n clones with highest affinities to be the antibodies in the new population $\mathbf{Ab}^{(t+1)}$ in the next iteration.
9. Finally, replace the $(1 - \frac{n}{N_{\mathbf{Ab}}}) \times 100\%$ lowest affinity antibodies in $\mathbf{Ab}^{(t)}$ by new individuals.
10. If not reaching the number of iterations, go back to step 3, otherwise, add the antibody with highest affinity, Ab_{i^*} to the population of the memory cells.
11. Set $k \leftarrow k + 1$ for the next component.

In step 3, the distance between each antibody Ab_i in $\mathbf{Ab}$ and the memory cells in $\mathbf{Ab}_{\{m\}}$ can be measured by the angle between two vectors,

$$D_{Ab_i} = \sum_{k=1}^{j-1} \left| \frac{Ab_i^{jT} \cdot Ab_{\{m\}}^k}{|Ab_i^j| \cdot |Ab_{\{m\}}^k|} \right|. \tag{10.6}$$

To illustrate our algorithm, we consider the problem of Canonical Correlation Analysis. We consider that there are two sets of antigens, $\mathbf{Ag}_1$ and $\mathbf{Ag}_2$ to represent the two sets of data and accordingly two populations of antibodies, $\mathbf{Ab}_1$ and $\mathbf{Ab}_2$, are generated to represent the two weight vectors. We define the affinity between the antigens and the antibody in form of

$$\mathbf{f}_i = \frac{1}{1 + \exp(\gamma |\mathbf{Ab}_{1i}^T \mathbf{X}_1 - \mathbf{Ab}_{2i}^T \mathbf{X}_2| + |(\mathbf{Ab}_{1i}^T \mathbf{X}_1)^2 - 1| + |(\mathbf{Ab}_{2i}^T \mathbf{X}_2)^2 - 1|)}. \tag{10.7}$$

where $\mathbf{X}_1 = \mathbf{Ag}_1$ and $\mathbf{X}_2 = \mathbf{Ag}_2$. We use the same artificial data set which was used in Subsection 8.1.3 and the size of data set is 10000. We set $N'_{\mathbf{Ab}} = 100, n = 10, \beta = 0.1, \alpha = 0.5, \gamma = 0.00001$ and for multiple components, $N_{\mathbf{Ab}} = 10000$. The number of iterations is 500.

We see in Table 10.6 that the correlation components (CCs) have been identified. Figure 10.5 and Figure 10.6 show the speed of convergence is fast. We also find that the convergence of immune-inspired algorithms is not as smooth as that of reinforcement learning and cross entropy; instead the algorithms can identify a local optimum quickly and finally converge to the global optimum.

Table 10.6. The first two Canonical Correlation (CC) weight vectors found with the artificial data

	m1	**m2**
	0.9967	**0.9999**
	-0.0510	0.0094
1st CC	0.0613	-0.0131
	0.0137	
	-0.0692	0.0028
	-0.9942	**-0.9986**
2nd CC	0.0146	0.0529
	0.0806	

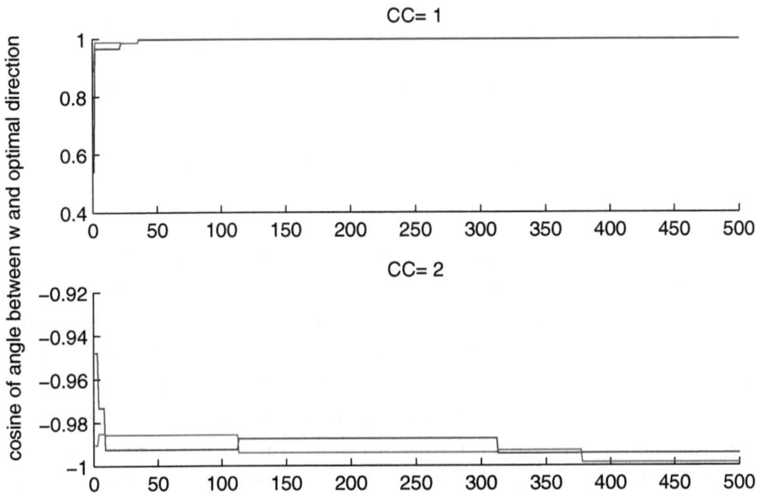

Fig. 10.5. Convergence of the CCA weight vectors to the optimal directions. Top: the 1st canonical correlation. Bottom: the 2nd canonical correlation.

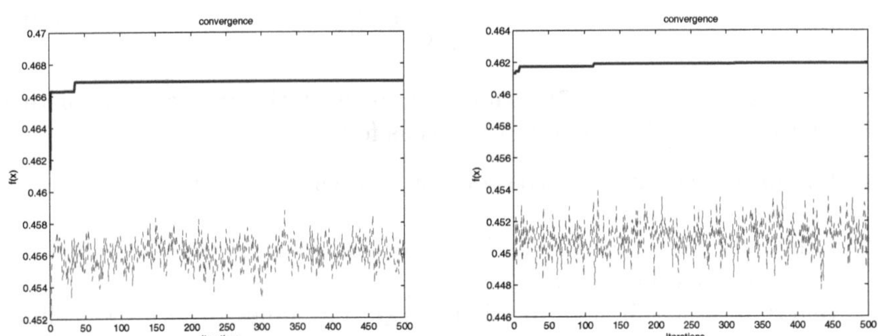

Fig. 10.6. Affinity of the population. Highest (solid line) and average (dashed line). Left: affinity for the 1st CC. Right: affinity for the 2nd CC.

10.3 Combining Adaptation Methods

In Chapter 8 and Chapter 9, we have extended reinforcement learning and cross entropy methods to solve projection problems. In the previous section of this chapter, we demonstrate that immune-inspired algorithms can be applied to solve the same problems. In this section, we develop algorithms that can achieve better performance in solving projection problems by combining these non-standard adaptation methods. The "better performance" is not only to increase the accuracy of the final optimized solution, but also decrease the size of data set needed for consistent performance and the number of iterations required. We first incorporate the cross entropy method into an artificial immune system, in which a new clone and maturation process is present. Then we show that the immune-inspired algorithm can be integrated with temporal difference learning, described in Subsection 8.3.1. The algorithms in this section are applied to identify the first principal component of PCA so that the results can be easily compared with those achieved previously.

10.3.1 Artificial Immune System with Cross Entropy

In this subsection, we incorporate the cross entropy method into an artificial immune system, and evaluate the optimal solution as that with highest accuracy. The cross entropy method can achieve the global optimum by defining a family of pdfs $\{f(.\,;\mathbf{v}), \mathbf{v} \in \mathcal{V}\}$ on the data set $\aleph$ and we make adaptive changes to the probability density function according to the Kullback-Leibler cross-entropy. Considering the modified CLONALG algorithm in Subsection 10.2.1, for a selected antibody $Ab_{\{n\}i}$ in population $\mathbf{Ab}_{\{n\}}$, we have a series of matured clones drawn from a Gaussian distribution with the mean $Ab_{\{n\}i}$ and the variance $\propto \exp(-\mathbf{f}_i)$. We denote the clones belonging to its parent antibody $Ab_{\{n\}i}$ as $\mathbf{C}_i^*$ with $\mathbf{c}_{ij}^*$ corresponding to the jth clone of the ith parent antibody. Thus we wish to maximize the affinity function $\mathbf{f}(\mathbf{c}_{ij}^*)$ over all clones in $\mathbf{C}_i^*$. Denoting the maximum by γ^*, we have

$$\gamma^* = \max_{\mathbf{c}_{ij}^* \in \mathbf{C}_i^*} \mathbf{f}(\mathbf{c}_{ij}^*) \tag{10.8}$$

Then we can associate (9.17) with (10.8), in which the cross entropy method is performed. The algorithm in detail is as follows

1. Initialize each antigen corresponding to each data point in the data set. The size of the data set is N.
2. Randomly generate a population of real-value antibodies $\mathbf{Ab}^{(t)}$. The size of the population is $N_{\mathbf{Ab}}$.
3. Determine the affinities to all the $N_{\mathbf{Ab}}$ antibodies to form the affinity vector $\mathbf{f}_i$.
4. Select the n antibodies with highest affinity from $\mathbf{Ab}^{(t)}$ to compose a new set $\mathbf{Ab}_{\{n\}}$.

5. Each selected antibody $Ab_{\{n\}i}$ is matured and cloned according to (10.4) to form a population $\mathbf{C}_i^*$. The number of clones in $\mathbf{C}_i^*$ is $n{\cdot}round(\beta{\cdot}N_{\mathbf{Ab}})$.
6. Determine the affinities $\mathbf{f}^*$ of the matured clones $\mathbf{C}_i^*$. Let $\hat{\gamma}_t$ be the $1-\varrho$ clone quantile, above which we identify the "elite" clones. Use the "elite" clones to generate a new antibody $Ab_i^{(t+1)}$ according to (9.24).
7. Put all these new antibodies $Ab_i^{(t+1)}, i = 1, \ldots, n$ to form part of the new population $\mathbf{Ab}^{(t+1)}$.
8. Replace the $(1 - \frac{n}{N_{\mathbf{Ab}}}) \times 100\%$ lowest affinity antibodies in $\mathbf{Ab}^{(t)}$ by new individuals to form the other part of $\mathbf{Ab}^{(t+1)}$.

Table 10.7. The first principal component with 5-dimensional artificial data by the modified CLONALG algorithmwith cross entropy

PC1	0.0036	0.0007	0.0077	-0.0053	**0.9999**

To illustrate the modified CLONALG algorithm with cross entropy, we use the same 5 dimensional data as in Subsection 8.1.2 in which the first principal component is readily identified as the fifth input dimension. The size of the data set is 10000 and we set $N_{\mathbf{Ab}} = 100, n = 10, \beta = 0.1, \alpha = 0.5, \varrho = 0.5$. The number of iterations is 100. We can see that our algorithm has identified the first principal component with much higher accuracy in Table 10.7. Figure 10.7 and Figure 10.8 show that our algorithm has converged

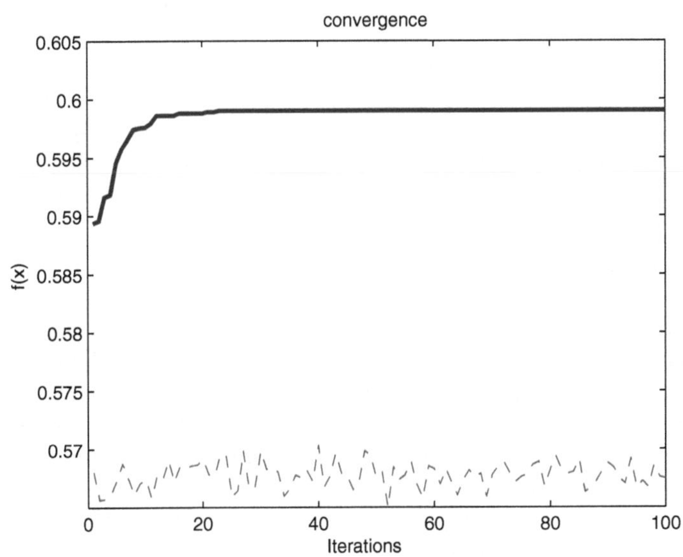

Fig. 10.7. Affinity of the population by the modified CLONALG algorithm. Highest (solid line) and average (dashed line).

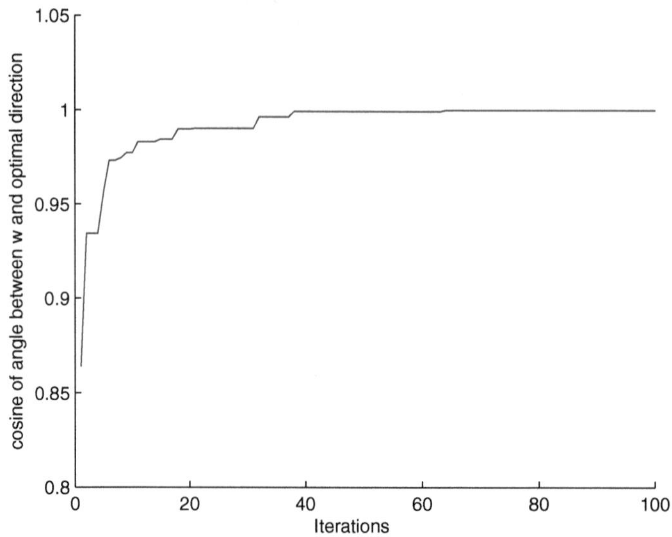

Fig. 10.8. Convergence of the PCA weight vector to the optimal directions. The vertical axis shows the cosine of the angle between the current filter and the optimal filter. The horizontal axis shows the number of iterations.

extremely fast. Furthermore, we have found that the cosine of the angle between the estimated weight vector and the optimal vector is higher than 0.99, even when the number of iterations is reduced to 10.

We compare the results by the classic CLONALG algorithm, the modified CLONALG algorithm and the CLONALG algorithm with cross entropy. The experiments are performed with the same size of data set, 10000, and the same number of iterations, 100. Figure 10.9 shows that although all of the algorithms can identify the first principal component quickly, the two latter algorithms are more efficient than the classic CLONALG algorithm.

10.3.2 TD Learning with Artificial Immune Systems

We have demonstrated that reinforcement learning with Q-learning (in Subsection 8.3.1) can be applied to solve projection problems with high accuracy. However, this algorithm suffers from the drawback that the speed of convergence is slow, which leads to a large number of iterations. In this subsection, we improve the algorithm described in 8.3.1 by integrating immune-inspired algorithms with Q-learning.

With reinforcement learning, the state of the system at any time is the data sample presented to the system at that time, i.e. $s_t = \mathbf{x}_t$. However, instead

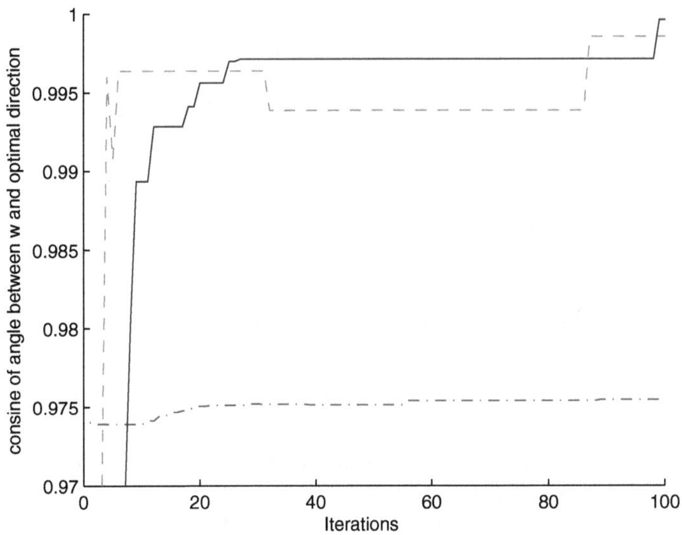

Fig. 10.9. Convergence of the PCA weight vector to the optimal directions. Dash-dot line: by the classic CLONALG algorithm. Dashed line: by the modified CLONALG algorithm. Solid line: by the CLONALG algorithm with cross entropy.

of sampling the weight vector $\mathbf{w}$ from the distribution $\mathcal{N}(\mathbf{m}, \beta^2 I)$ with the current estimate of the parameters, $\mathbf{m}$ and β^2, we define the action to be to generate a population of antibodies Ab_i, where the modified CLONALG algorithm is performed to maximize the Q-value of the current data point $\mathbf{x}_t$. Thus the affinity function is defined to calculate the Q-value of the antibody. At each iteration, we keep a note of the greatest Q-value and only update the Q-value of the data point by the antibody Ab_{i*} with the highest Q-value. We therefore re-structure the learning algorithm as follows

1. Randomly generate a population of real-value antibodies $\mathbf{Ab}^{(t)}$. The size of the population is $N_{\mathbf{Ab}}$.
2. Randomly select a data point $\mathbf{x}_t$ to be the current state s_t.
3. Determine the reward r_i for each antibody Ab_i in population $\mathbf{Ab}$.
4. Determine the Q-value of each antibody Ab_i in population $\mathbf{Ab}$ to form the affinity vector $\mathbf{f}_i$ according to (8.22) and (8.24).
5. Select the n antibodies with highest affinity from $\mathbf{Ab}$ to compose a new set $\mathbf{Ab}_{\{n\}}$. Denote the antibody with the highest affinity as Ab_{i*}.
6. Update the Q-value of data point $\mathbf{x}_t$ with the affinity of Ab_{i*}.
7. Mature and clone the n selected antibodies independently according to (10.4) to form a population $\mathbf{C}^*$. The size of population of clones is $N_c = \sum_{i=1}^{n} round(\beta \cdot N_{\mathbf{Ab}}) = n \cdot round(\beta \cdot N_{\mathbf{Ab}})$.
8. Determine the affinity $\mathbf{f}^*$ of the matured clones $\mathbf{C}^*$.

9. From this population of mature clones $\mathbf{C}^*$, select n clones with highest affinities to be the antibodies in the new population $\mathbf{Ab}^{(t+1)}$ in the next iteration.
10. Finally, replace the $(1 - \frac{n}{N_{\mathbf{Ab}}}) \times 100\%$ lowest affinity antibodies in $\mathbf{Ab}^{(t)}$ with new individuals.

Table 10.8. The first principal component with 5-dimensional artificial data by Q-learning with immune-inspired algorithm

PC1	-0.0194	0.0125	0.0647	0.1152	**-0.9910**

To illustrate the Q-learning with immune-inspired algorithm, we use the same 5 dimensional data in Subsection 8.1.2 in which the first principal component is readily identified as the fifth input dimension. The size of the data set is 100 and we set $N_{\mathbf{Ab}} = 100, n = 10, \beta = 0.1, \alpha = 0.5$. The number of iterations is 2000. We can see that our algorithm has identified the first principal component with high accuracy in Table 10.8. Figure 10.11 shows that our algorithm converges quickly and the greatest Q-value is maximized step by step as shown in Figure 10.10.

We can see that our Q-learning with immune-inspired algorithm has improved the performance of the Q-learning method in projection problems in terms of the number of iterations and with the size of data set extremely reduced. Figure 10.11 also shows that when the number of iterations is 2000

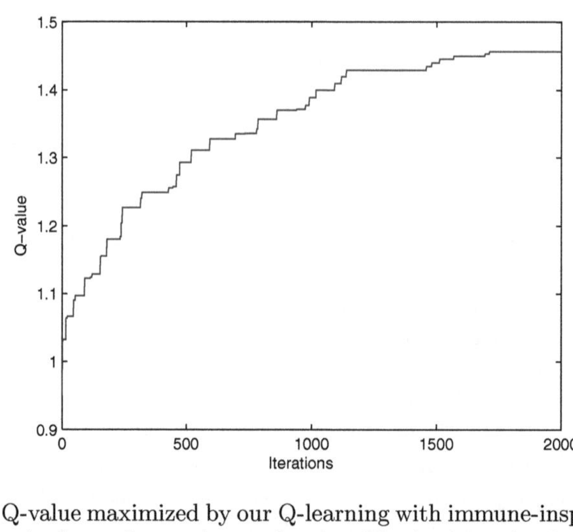

Fig. 10.10. Q-value maximized by our Q-learning with immune-inspired algorithm. The vertical axis shows the maximum Q-value. The horizontal axis shows the number of iterations.

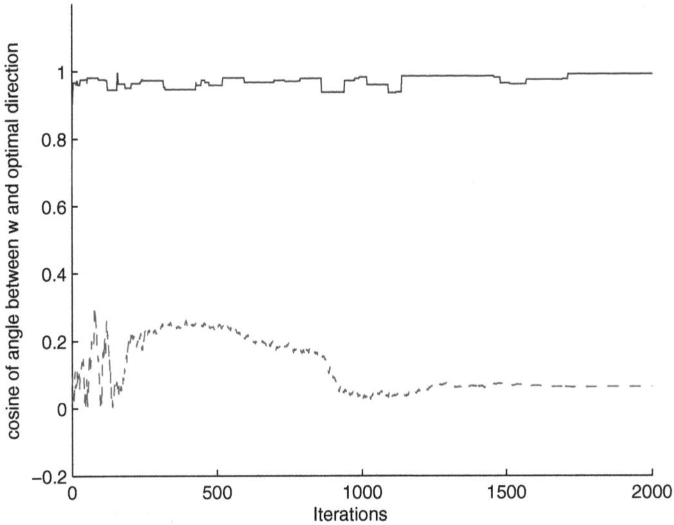

Fig. 10.11. Convergence of the PCA weight vector to the optimal directions. Dashed line: by Q-learning method in Subsection 8.3.1. Solid line: by Q-learning with immune-inspired algorithm in this subsection.

and the size of data set is 100, the Q-learning method individually performed in Subsection 8.3.1 can not identify the first principal component: we have to increase the size of the data set and the number of iterations so that this Q-learning method can converge to the optimal solution.

10.4 Ensembles of the Non-standard Adaptation Methods

In the previous chapters, algorithms based on non-standard adaptation methods have been developed to solve projection problems and the results have shown that all the algorithms can converge to the optimum stably and with high accuracy. However, to achieve highly accurate results, it is always necessary to set the parameters of an algorithm carefully and usually a high volume of data is required, otherwise, the algorithm may fail to find the optimal solution, or become unstable. In this section, we investigate ensemble methods in the context of using non-standard adaptation methods to solve projection problems. Specifically, we investigate bagging to perform reinforcement learning, cross entropy methods and immune-based algorithms in parallel, through which the converged optimum is more stable and reliable and with high accuracy.

10.4.1 Bootstrapping and Bagging

Bagging was proposed by Breiman [40], which is based on bootstrapping [74] and aggregating concepts. Bootstrapping [74] is a simple and effective way of estimating a statistic of a data set. Suppose we have a data set, $D = \mathbf{x}_i, i = 1, \ldots, N$: bootstrapping creates a number of pseudo data sets, D_i, by sampling from D with uniform probability with replacement of each sample. Thus each data point has a probability of $\left(\frac{N-1}{N}\right)^N \approx 0.368$ of not appearing in each bootstrap sample, D_i. Aggregating means we can perform the same or even different algorithms in parallel. Typical applications with bagging are prediction [41, 72], classification [206] and data pre-processing [232].

In detail, bagging consists of selecting B data sets, $D_1, \ldots, D_B$ from D by randomly selecting a member of D with replacement and each data set D_b will almost certainly contain only some members of the original data set D. From a classification perspective, each classifier $C^b(\mathbf{x})$ is then constructed based on the data set D_b. All the classifiers are combined by simple majority voting or some average combination rule. Furthermore, the elements of $T_b = D - D_b$ can be used to assess how accurate the training of the individual classifier is liable to be.

10.4.2 Non-standard Adaptation Methods with Bagging

We [232].consider the problem of principal component analysis and use the same 5 dimensional data as in Subsection 8.1.2 in which the first principal component is readily identified as the fifth input dimension. The size of data set is 1000. We create the first 20 bags by selecting from the data set and we perform the reinforcement learning algorithm derived in Subsection 8.3.1 on each bag. The number of iterations is 200000. The learning rate is initialized to 0.1, which is reduced step by step. We create the second 20 bags using the cross entropy method with batch-learning from Subsection 9.3. The number of iterations is 2000. We set the number of random samples as each iteration as 50 and $\varrho = 0.1$. We create the last 20 bags with the modified CLONALG algorithm in Subsection 10.2.1. The number of iterations is 500 and we set $N_{\mathbf{Ab}} = 100, n = 10, \beta = 0.1, \alpha = 0.5$. For all of the 60 bags, the number of data points in each bag, N_{D_b} is 500.

Therefore, each bag has one *local solution*, $\mathbf{w}_b$. We present a new way, *self-organized majority voting*, to decide the final solution according to these local solutions. We first evaluate how well each local solution has been optimized by

$$J_{\mathbf{w}_b} = \sum_{i=1}^{N} |\mathbf{w}_b \mathbf{x}_i|. \tag{10.9}$$

We denote that local solution that has been optimized best as the winning local solution , $\mathbf{w}_{b^*}$. Then the weight of one local solution in the final solution is defined proportional to the distance between each local solution, $\mathbf{w}_b$, and the winner, $\mathbf{w}_{b^*}$,

$$h_{b^*}(b) = \exp(-\gamma ||\mathbf{w}_b - \mathbf{w}_{b^*}||^2) \tag{10.10}$$

Then the final solution is calculated by

$$\mathbf{w} = \frac{\sum_{b=1}^{B} h_{b^*}(b)\mathbf{w}_b}{\sum_{b=1}^{B} h_{b^*}(b)}. \tag{10.11}$$

We compare the final solution by our method with the average first principal component filters by reinforcement learning, cross entropy method and the modified CLONALG algorithm respectively. Table 10.9 shows that the final solution by this method has higher accuracy than the individual methods. We can also see that bagging with self-organized majority voting is more effective than bagging with simple majority voting.

Table 10.9. The first principal component filters identified by bagging with self-organized majority voting and simple majority voting and the average first principal component filters by reinforcement learning, cross entropy method and the modified CLONALG algorithm

Self-organized voting	0.0143	0.0197	0.0501	0.0915	**0.9943**
Simple voting	0.0142	0.0197	0.0505	0.0957	**0.9909**
Reinforcement Learning	0.0136	0.0149	0.0400	0.0887	**0.9929**
Cross Entropy Method	0.0103	0.0192	0.0519	0.1176	**0.9877**
CLONALG algorithm	0.0188	0.0249	0.0599	0.0814	**0.9920**

We decrease the size of data set to 500 with 100 data points in each bag. In such a situation, all the non-standard adaptation algorithms used in this subsection become much more unstable and unreliable. However, we see that the final solution by bagging with self-organized majority voting is much more stable and reliable than those by other methods as shown in Table 10.10.

Table 10.10. The first principal component filters identified by bagging with self-organized majority voting and simple majority voting and the average first principal component filters by reinforcement learning, cross entropy method and the modified CLONALG algorithm. The size of data set is decreased to 100.

Self-organized voting	0.0198	0.0369	0.0707	0.1086	**0.9907**
Simple voting	0.0224	0.0461	0.0978	0.3109	**0.8974**
Reinforcement Learning	0.0217	0.0380	0.0991	0.2126	**0.9531**
Cross Entropy Method	0.0247	0.0558	0.1047	0.4664	**0.8018**
CLONALG algorithm	0.0208	0.0456	0.0910	0.2770	**0.9218**

10.5 Conclusion

We have, in this chapter, extended the artificial immune system to solve projection problems by developing immune-inspired algorithms. We have first demonstrated that the CLONALG algorithm can be applied to independent component analysis. The data set is represented by antigens and the artificial immune system improves the affinity by generating a series of populations of antibodies. The affinity is defined in different ways in order to solve different projection problems. We have performed the affinity maturation process used by artificial immune networks in the CLONALG algorithm, which we consider to be more suitable for projection problems. We have also pointed out that giving a smaller variance in the cloning and maturation process can greatly improve the performance.

We have also presented a new way to perform deflationary orthogonolization, in which multiple components are identified by the immune system directly. We regard the weight vectors previously estimated as memory cells and the basic idea is that the distance between antibodies selected to be cloned and matured in a new population and those memory cells, is as large as possible. Thus for a population of antibodies, we select the antibodies with largest distance to the memory cells, based on which we re-select the antibodies with highest affinity. We consider this way to be particularly suitable for an artificial immune system in that all components can be identified with the same affinity function.

We have investigated the combination of different non-standard adaptation methods. We first incorporated the cross entropy method into the modified CLONALG algorithm. Rather than only selecting that clone with highest affinity from the set of clones belonging to the same antibody as the antibody in the new population, we take all the clones belonging to the same antibody into account and generate the antibody in the new population according to (9.24) and (9.25), which is the same thing the cross entropy method does. The results show that this combined algorithm has higher accuracy with faster convergence. We have compared it with the modified CLONALG algorithm and the modified CLONALG algorithm. Then we integrated the modified CLONALG algorithm with the Q-learning method. All the antibodies aim to maximize the Q-value of the current state. We demonstrated that such a combined algorithm can not only identify the principal component with high accuracy but also can do so with a smaller size of data set. The number of iterations has also been extremely reduced.

We have applied our algorithms to ICA, PCA and CCA as examples. Especially we have only demonstrated the combined algorithms in identifying the 1st principal component, which is more convenient to compare the performance of different algorithms. However, it is worth noting that all the immune-inspired algorithms are general methods that can be applied to the other projection problems.

Finally, we have combined the three non-standard adaptation methods discussed in this book with bagging. We have also presented a self-organized majority voting method to decide the final solution based on the local solutions, where the proportion of a local solution in the final solution is decided by the distance between that local solution and the winner solution. The experiments have shown our method can make the algorithms more stable and reliable and with higher accuracy.

11

Conclusions

We have introduced several non-standard adaptation methods in this book and we now review the book and point out possible avenues for future work. However we first wish to address the question why we use these methods on very standard exploratory data analysis techniques.

11.1 Rationale

Thus, at the end of this book, we will address the question why we need to develop additional learning mechanisms for these tasks when, after all, the tasks we have investigated in this book have all been addressed (and perhaps more efficiently) by other means. This argument has previously been used with respect to neural networks particularly those performing PCA. One answer clearly, as we have seen with neural networks, is that the experience gained on such models was invaluable in creating non-linear PCA. Our motivation for investigating these non-standard adaptation methods comes from two sources:

1. Although, for example, PCA can be performed efficiently by standard methods, alternative algorithms intended to robustify PCA cannot always be efficiently created. For example, [70] discusses a form of PCA which uses the L_2 norm in data space while utilising the L_1 norm when considering the error over the whole data set. Thus we have, for PCA, to minimise

$$J_{PCA} = \sum_{i=1}^{N} \| \mathbf{x}_i - \hat{\mathbf{x}}_i \|^2 \tag{11.1}$$

while the robust version seeks to minimise

$$J_{R_1-PCA} = \sum_{i=1}^{N} \| \mathbf{x}_i - \hat{\mathbf{x}}_i \| \tag{11.2}$$

W.A. Barbakh et al.: Non-Stand. Param. Adapt. for Explor. Data Analys., SCI 249, pp. 199–205.
springerlink.com

for which an iterative algorithm has to be developed. However, the cost function is differentiable and so any of the methods used in this book will perform this minimisation. Thus, for example, we have used immediate reward reinforcement learning on this cost function and Figure 11.1 shows the convergence of the first principal component filter of the artificial data set in Subsection 8.1.2. We set the size of the data set to 1000 and the number of iterations is 50000. We concur with [70] that it is more robust to large amplitude shot noise than standard PCA and the number of iterations is much less than for standard PCA.

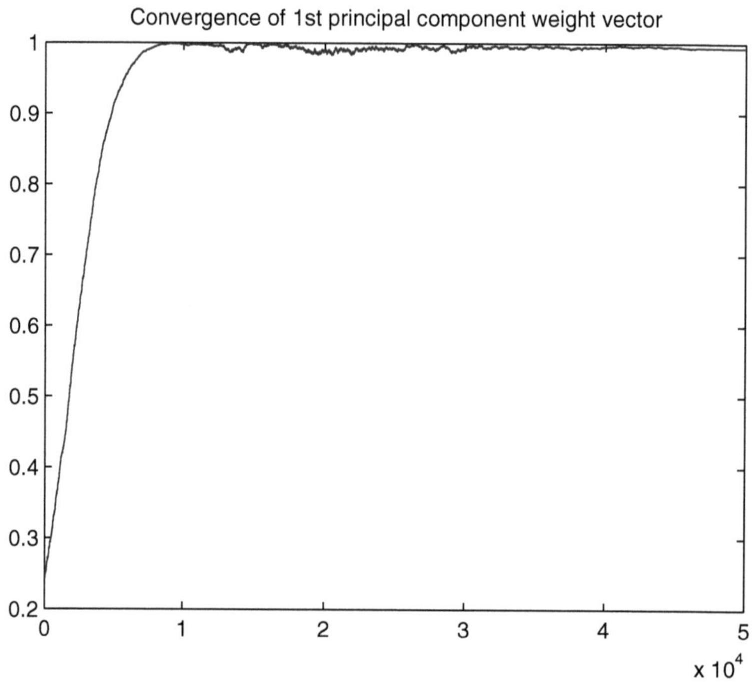

Fig. 11.1. Convergence of the first PCA weight vectors to the optimal directions by R_1-PCA. The vertical axis shows the cosine of the angle between the current filter and the optimal filter. The horizontal axis shows the number of iterations.

2. There has been a great deal of recent interest in "deep architectures" [27, 114, 112, 25]. This has stemmed from a dissatisfaction with current methods for supervised training, for example, of classifiers. Deep learning substitutes many layers, each a narrow layer of neurons, for one or two broad layers of neurons. It is stated that such networks are more efficient: for example, it is well known that standard backpropagation may often get stuck in local minima and this problem is known to be exacerbated when using multiple layers of neurons. Hinton [114] shows that

pre-processing a data set by training several layers of Restricted Boltzmann Machines on a data set and then using the resulting network as the starting point for supervised (backpropagation) learning, results in a very much more efficient learning machine. Thus the search is on for alternative learning architectures and methods. We have used a Boltzmann-based reinforcement learning [196] on the XOR problem and shown that it converges reliably to solve this problem; however more research into more complex problems which require deeper architectures is necessary.

Then we may ask why we have chosen to discuss the methods in this book utilised on simple tasks like PCA, ICA, CCA etc. The advantage that these simple functions have is that their optima are easily demonstrated. Thus we have been mainly involved in comparing these algorithms on these simple problems as one can easily see how effective the respective algorithms are. With more complex problems the answers may not be so obvious as with these simple problems. We deviated from this once in Chapter 9, in which we used cross-entropy on a deep architecture structure. The results were interesting but it is not obvious how to compare e.g. our results in Chapter 9 with multiple layers of standard GTMs whereas with PCA etc., the comparison is readily performed.

11.2 Summary and Remarks

We began, in this book, by investigating algorithms which improve upon K-means, one of the most popular clustering algorithms. The major deficiency in K-means is its sensitivity to initial conditions which leads to convergence to local optima rather than the global optimum. This deficiency is not unique to K-means; for example, the mixture of experts algorithm trained by Expectation-Maximization (the EM algorithm) which is another very popular clustering method, is also known to suffer from this defect. We consider that it is the local nature of these algorithms which is at the heart of this deficiency and have sought to introduce global knowledge into the method of placing the prototypes. Mathematically, for both of the named existing algorithms, we have (with the notation of previous chapters) $\frac{\partial \mathbf{m}_k}{\partial \mathbf{x}_i} = 0$ for $\mathbf{x}_i$ not sufficiently local to $\mathbf{m}_k$. This leads to computationally efficient algorithms (the tactic of divide and conquer is a well-tried technique) but can not guarantee to find the global optimum. A similar argument can be made in other fields such as path-finding: consider the Traveling Salesman Problem in which a salesman has to visit a number of cities in an efficient path without visiting each city more than once. A human response to this might be to start making local sections of the route while keeping in mind global relationships between the local sections. We wish to do something similar with our algorithms for clustering: make the local clusters the driving force for positioning the prototypes but ensure that global knowledge is also used to drive the overall positioning of the set of prototypes. Mathematically our algorithms exhibit $\frac{\partial \mathbf{m}_k}{\partial \mathbf{x}_i} \neq 0, \forall k, \forall i$.

However we can be a little more precise than that because our algorithms also tend to have $\frac{\partial \mathbf{m}_k}{\partial \mathbf{x}_i}$ greatest for data points, $\mathbf{x}_i$, local to $\mathbf{m}_k$ so that the driving force of the local algorithms is retained as the major imperative (K-means has after all been a very successful algorithm) while incorporating the global effect which we have shown to be necessary. It is this blend of local and global information which has been the basis of our algorithms' success. We believe that we have shown this tactic to be successful empirically but note that this remains an heuristic (just as many of the best current solutions to the Traveling Salesman Problem are based on heuristics).

We also extend reinforcement learning to unsupervised data exploration, in particular, to finding valid projections. Three general reinforcement learning methods have been derived for projection problems under the framework of reinforcement learning. We have presented a particular structure in that the component weight vectors are represented by stochastic units drawn from the Gaussian distribution with mean $\mathbf{m}$ and variance $\beta^2 I$ and we take adaptive update rules for the parameters of the Gaussian distribution so as to maximize the expected value of the long-term reward. We have demonstrated our immediate reward reinforcement learning on independent component analysis, where a set of reward functions has been developed from different perspectives. Then we have shown that the method is such a general method that it can easily be applied to other projection problems. In addition, we have also shown that the reward function can be extended to contain multiple factors with different meanings so that the reinforcement learning is constrained with more than one condition. In doing so, we have combined deflationary orthogonalization directly into the reward function so that the action is only decided by the reward function and we have very fast convergence with high accuracy.

We have described an alternative immediate reward reinforcement learning in which the individual unit is deterministic but has stochastic synapses. Four learning rules that have been used in a supervised learning neural network have been applied to optimize the distribution of the weight vectors. The results shows that the method can converge smoothly with these learning rules but they have different efficiencies.

Considering that allocating incremental rewards to actions do in fact lead to the optimal reward in the long run, we have used two reinforcement learning algorithms based on temporal difference learning, one of which is based on Q-learning and the other of which is based on Sarsa-learning. We have applied the algorithm based on Q-learning to linear projection and the algorithm based on Sarsa-learning to kernel projections. We have shown that we have fast convergence with accurate and stable results. Additionally, for kernel projections, we have developed a different way to identify multiple directions in the feature space.

In Chapter 9, we have extended the cross entropy method to solve projection problems. The projections have been considered as optimization problems that can be turned into rare-event estimation and then the cross entropy method is performed with a variety of appropriate performance functions. We have used the same stochastic units to sample the weight vectors and with the cross entropy method, we make adaptive changes to the probability density function of component weight vectors according to the Kullback-Leibler cross-entropy. Apart from the batch cross entropy method, we have derived an on-line cross entropy method, in which the adaptive update of the parameters of the Gaussian distribution is smoother. We also applied the general method to identify multiple components in the same sort of way which has used in reinforcement learning. The results have shown that we have very fast and stable convergence with high accuracy. We also extended the cross entropy method to latent variable models, where we perform the optimization of latent variable models as associated stochastic problems with batch or on-line cross entropy methods.We finally developed a multilayer topology preserving mapping for clustering, by which we demonstrated deep architecture in unsupervised data exploration. The results have shown the model identified more abstract features in higher layers, in which the clusters of the data set are identified accurately.

In Chapter 10, we have extended artificial immune systems to solve projection problems. First, we have demonstrated that a modified CLONALG algorithm can be applied to linear projections directly, where the data set is represented by antigens and the artificial immune system improves its affinity by generating a series of populations of antibodies. We have pointed out that the algorithm converges faster and more stably with higher accuracy with a smaller variance in the cloning and maturation process. We have continued to extend the modified CLONALG algorithm to identify multiple components. We regard the weight vectors previously estimated as memory cells and only those antibodies with largest distance from the memory cells are selected to be cloned and matured in a new population. We have shown that the new algorithm can identify the components accurately.

We have improved the performance of the non-standard adaptation algorithms by combining the algorithms. We have pointed out that the "better performance" is not limited to improving the accuracy of the final results, but also reduces the size of data set required and the number of iterations. To achieve the former, we have incorporated the cross entropy method into the immune system. We have compared the results from the combined algorithm with that from the modified CLONALG algorithm. We have achieved the latter by integrating the immune-inspired algorithm with the Q-learning method. We have shown that both the size of data set and number of iterations have been extremely decreased.

We have finally developed a bootstrapping method to perform reinforcement learning, the cross entropy method and the immune-inspired algorithms in parallel. We have derived a new way to decide the final solution, called *self-organized majority voting*, which we consider to be more suitable for projections that simple majority voting. We thus have more stable and reliable solutions for projections.

11.3 Further Research

Although we have shown that the immediate reward reinforcement learning methods can converge fast and stably, the value of the parameters of the methods must be set accurately. For example, if the learning rate in the immediate reward learning is too small and the number of iterations is not large enough, the algorithm will not converge. Of course, we can either increase the learning rate or increase the number of iterations. However, if we increase the learning rate, the algorithm will not perform as smoothly as before, and on the other hand, increasing the number of iterations leads to longer running-times and heavier volume of calculations. Such situations happen especially when defining more than one factor in the reward function. Therefore, the more parameters there are in the algorithm, the more attention we have to pay to setting the value of the parameters. It would be useful to determine optimal parameters analytically.

We have introduced the cross entropy method as a powerful approach to solving rare-event problems. We make adaptive changes to the probability density function of component weight vectors according to the Kullback-Leibler cross-entropy to solve projection problems. It may be interesting to investigate other entropies such as Tsallis or Renyi entropies.

We have demonstrated that we have extremely fast convergence and high accuracy using immune-inspired algorithms. Instead of using stochastic units to sample the weight vectors, the antibodies in the artificial immune system form a search space, in which the global optimum can be discovered step by step. In doing so, the precondition is that we assume the global or local optimum is in the search space. However, if the problem becomes too complicated, the algorithms derived in this book can not guarantee to find the global optimum. A simple example is when we apply the extended CLONALG algorithm in Subsection 10.2.2 to identify the fifth projection of the 5-dimensional data set in Subsection 8.1.2. The projection problem becomes more and more complicated from identifying the first principal component vector to identifying the fifth principal component vector. Therefore, the number of the antibodies must be increased from 100 for the 1st principal component to 2000000 for the 5th. In effect we are enlarging the search space. In Table 11.1, we can see that although the algorithm has identified the first few principal component filters of the data set, however, the last principal component filter is less accurately estimated, which obviously needs to be improved.

The three non-standard adaptation methods in this book require smaller size of data set compared to some other widely used algorithms, however they

Table 11.1. The weights from the artificial data experiment for five principal components with immune-inspired algorithm described in Subsection 10.2.2

PC1	0.0044	-0.0205	-0.0035	0.0877	**-0.9949**
PC2	-0.0553	0.0054	0.0355	**0.9915**	0.0532
PC3	0.0078	0.0151	**0.9958**	-0.0422	0.0286
PC4	0.0055	**-0.9906**	-0.0786	0.0760	0.0930
PC5	**0.9560**	0.546	0.0435	0.0930	0.1476

are sensitive to the size of data set, which means it will be difficult to reach the global optimum for these methods if not given enough data. Although the combined algorithm in Subsection 10.3.2 has decreased the size of data set required and we have a more stable and reliable final solution by using bagging, it is still a topic for future reseach.

In this book, we solve the projection problems by optimizing the weight vectors by non-standard adaptation methods. However, we keep in mind that the task of projection is to obtain data points in lower dimensions so that we can perform clustering or visualization more easily. Evaluation of the data points in projection space directly has already been done using bayesian methods, such as GPLVM-PCA [146] or GPLVM-CCA [147]. This is worth future research in the context of non-standard adaptation methods.

Non-linear projection problems are more difficult and complex tasks than linear projections. We have applied the algorithms we have derived to kernel projections and latent variable models, the utility of these methods in the context of other non-linear projection problems could be investigated in future reseach.

We believe that this book has made contributions to the research in the field of exploratory data analysis but clearly there are still many avenues of research to pursue.

References

1. Data Mining: Concepts and Techniques (2007), `http://www.cs.sfu.ca/~han/bk/8clst.ppt#1`
2. Ackley, D.H., Littman, M.L.: Generalization and scaling in reinforcement learning. In: Advances in Neural Information Processing Systems, vol. 2, pp. 550–557 (1990)
3. Ada, G.L., Nossal, G.J.V.: The clonal selection theory. Scientific American 257(2), 50–57 (1987)
4. Almeida, L.B.: MISEP- linear and nonlinear ica based on mutual information. Journal of Machine learning Research 4, 1297–1318 (2003)
5. Anders, P.: Using temporal difference methods in combination with artificial neural networks to solve strategic control problems. Master's thesis, Department of Numerical Analysis and Computer Science, Royal Institute of Technology (2004)
6. Anderson, E.: The irises of the gaspe peninsula. Bulletin of the American Iris Society 59, 2–5 (1935)
7. Anderson, E.C.: Monte carlo methods and importance sampling. Lecture Notes for Statistical Genetics (1999)
8. Anderson, T.W.: Asymptotic theory for principal component analysis. Annals of Mathematical Statistics 34, 122–148 (1963)
9. Arthur, D., Vassilvitskii, S.: K-means++: The advantages of careful seeding. In: Bay Area Theory Symposium, BATS 2006 (2006), `http://www.stanford.edu/~sergeiv/papers/kMeansPP-soda.pdf`
10. Bach, F.R., Jordan, M.I.: Kernel independent component analysis. Journal of Machine Learning Research 3, 1–48 (2002)
11. Barbakh, W.: The family of inverse exponential k-means algorithms. Computing and Information Systems 11(1), 1–10 (2007)
12. Barbakh, W.: Similarity graphs. Computing and Information Systems 11(3), 1–11 (2007)
13. Barbakh, W., Crowe, M., Fyfe, C.: A family of novel clustering algorithms. In: Corchado, E., Yin, H., Botti, V., Fyfe, C. (eds.) IDEAL 2006. LNCS, vol. 4224, pp. 283–290. Springer, Heidelberg (2006)
14. Barbakh, W., Fyfe, C.: Performance functions and clustering algorithms. Computing and Information Systems 10(2), 2–8 (2006)

15. Barbakh, W., Fyfe, C.: Clustering with reinforcement learning. In: Yin, H., Tino, P., Corchado, E., Byrne, W., Yao, X. (eds.) IDEAL 2007. LNCS, vol. 4881, pp. 507–516. Springer, Heidelberg (2007)
16. Barbakh, W., Fyfe, C.: Inverse weighted clustering algorithm. Computing and Information Systems 11(2), 10–18 (2007)
17. Barbakh, W., Fyfe, C.: Tailoring local and global interactions in clustering algorithms. Technical Report 40, School of Computing, University of the West of Scotland, University of Paisley, UK (March 2007) ISSN 1461-6122
18. Barbakh, W., Fyfe, C.: Clustering and visualization with alternative similarity functions. In: The 7th WSEAS international conference on artificial intelligence, knowledge, engineering and data bases, AIKED 2008, University of Cambridge, UK, February 2008, pp. 238–244 (2008) ISSN: 1790-5109 ISBN: 978-960-6766-41-1
19. Barbakh, W., Fyfe, C.: Local vs global interactions in clustering algorithms: Advances over k-means. International Journal of Knowledge-based and Intelligent Engineering Systems 12 (2008) ISSN 1327-2314
20. Bartholomew, D.J.: Latent Variable Models and Factor Analysis. Charles Griffin Co. Ltd., London (1987)
21. Basilevsky, A.: Statistical Factor Analysis and Related Methods. Wiley, Chichester (1994)
22. Bell, A., Sejnowski, T.: Learning the higher-order structure of natural sound. Network: Computation in Neural Systems 7, 261–266 (1996)
23. Bellman, R.E.: Dynamic Programming. Princeton University Press, Princeton (1957)
24. Bellman, R.E.: A markov decision process. Journal of Mathematical Mechanics 6, 679–684 (1957)
25. Bengio, Y., Erhan, D., Courville, A., Larochelle. H., Bergstra, J.: An empirical evalution of deep architectures on problems with many factors of variation. In: Proceedings of the International Coference on Machine Learning, 2007, pp. 473–480 (2007)
26. Bengio, Y., Lamblin, P., Popovici, D., Larochelle, H.: Greedy layer-wise training of deep networks. In: Advances in Neural Information Processing Systems, vol. 19, pp. 153–160. MIT Press, Cambridge (2007)
27. Bengio, Y., LeCun, Y.: Large-Scale Kernel Machines. In: Scaling Learning Algorithms towards AI. MIT Press, Cambridge (2007)
28. Berkhin, P.: Survey of clustering data mining techniques. Technical report, Accrue Software (2002)
29. Berkhing, P.: Survey of clustering data mining techniques. Technical report, Accrue Software (2002), `http://www.ee.ucr.edu/~barth/EE242/clustering_survey.pdf`
30. Bertsekas, D.P., Tsitsiklis, J.N.: Neuro-dynamic programming. Athena Scientific (1996)
31. Bezedek, J.C.: A convergence theorem for the fuzzy isodata clustering algorithms. IEEE Transactions Pattern Analysis and Machine Intelligence 2(1), 1–7 (1980)
32. Bezedek, J.C.: Pattern recognition with fuzzy objective function algorithms. Plenum Press, New York (1981)
33. Bishop, C.M.: Latent variable models. In: Learning in Graphical Models, pp. 371–403. MIT Press, Cambridge (1999)

34. Bishop, C.M., Svensen, M., Williams, C.K.I.: GTM: The generative topographic mapping. Neural Computation 10(1), 215–234 (1997)
35. Bishop, C.M., Svensen, M., Williams, C.K.I.: Magnification factors for the gtm algorithm. In: Proceeding of the IEE 5th International Conference on Artificial Neural Networks, Cambridge, UK, pp. 64–69 (1997)
36. Bishop, C.M., Svensen, M., Williams, C.K.I.: Developments of the generative topographic mapping. Neurocomputing 21(1), 203–224 (1998)
37. Bishop, C.M., Svensen, M., Williams, C.K.I.: GTM: The generative topographic mapping. Neural Computation 10, 215–234 (1998)
38. Bishop, C.M.: Pattern Recognition and Machine Learning. Springer, Heidelberg (2006)
39. Boser, B.E., Guyon, I.M., Vapnik, V.N.: A training algorithm for optimal margin classifiers. In: 5th Annual ACM Workshop on COLT, pp. 144–152. ACM Press, New York (1992)
40. Breiman, L.: Bagging predictors. Machine Learning 24(2), 123–140 (1996)
41. Breiman, L.: Using updaptive bagging to debias regression. Technical Report 547, Statistics Department, University of California (1999)
42. Bronstein, A.M., Bronstein, M.M., Kimmel, R.: Generalized multidimensional scaling: a framework for isometry-invariant partial surface matching. Proceeding of National Academy of Sciences 103, 1168–1172 (2006)
43. Burges, C.: A tutorial on support vector machines for pattern recognition. Data Mining and Knowledge Discovery 2(2), 121–167 (1998)
44. Burnet, F.M.: The Clonal Selection Theory of Acquired Immunity. Cambridge University Press, Cambridge (1959)
45. Burnet, F.M.: Clonal selection and after. Theoretical Immunology 300(19), 1105–1107 (1978)
46. Campelo, F., Watanabe, K., Igarashi, H.: 3d topology optimization using an immune algorithm. The International Journal for Computation and Mathematics in Electrical and Electronic Engineering 26(3), 677–688 (2006)
47. Carter, J.H.: The immune system as a model for pattern recognition and classification. Journal of the American Medical Informatics Association 7(1), 28–41 (2000)
48. Chakrabarti, S.: Mining the Web: Discovering Knowledge from Hypertext Data. Morgan Kaufmann, San Francisco (2003)
49. Cichocki, A.A., Yang, H.H.: A new learning algorithm for blind signal separation. In: Advances in Neural Information Processing Systems, vol. 8, pp. 757–763 (1996)
50. Coello-Coello, C., Cruz-Cortes, N.: An approach to solve multiobjective optimization problems based on an artificial immune system. In: First International Conference on Artificial Immune Systems (ICARIS-2002), pp. 212–221 (2002)
51. Comon, P.: Independent component analysis - a new concept? Signal Processing 36, 287–314 (1994)
52. Cristianini, N., Shawe-Taylor, J.: An Introduction to Support Vector Machines and other kernel-based learning methods. Cambridge University Press, Cambridge (2000)
53. Crites, R.H., Barto, A.G.: Improving elevator performance using reinforcement learning. In: Advances in Neural Information Processing System, vol. 8, pp. 1017–1023 (1996)

54. Cutello, V., Narzisi, G., Nicosia, G., Pavone, M.: Clonal selection algorithms: A comparative case study using effective mutation potentials. In: Jacob, C., Pilat, M.L., Bentley, P.J., Timmis, J.I. (eds.) ICARIS 2005. LNCS, vol. 3627, pp. 13–28. Springer, Heidelberg (2005)
55. Cutello, V., Nicosia, G., Oliveto, P.S., Romeo, M.: On the convergence of immune algorithms. In: Proceedings of the first IEEE symposium on Foundations of Computational Intelligence, Hawaii, USA, pp. 409–415 (2007)
56. de Boer, P.-T., Kroese, D.P., Mannor, S., Rubenstein, R.Y.: A tutorial on the cross-entropy method. Annals of Operations Research 134(1), 19–67 (2004)
57. de Boer, P.T., Kroese, D.P., Rubinstein, R.Y.: Estimating buffer overflows in three stages using cross-entropy. In: Proceedings of the 2002 Winter Simulation Conference, pp. 301–309 (2002)
58. de Castro, L.N., Timmis, J.I.: Artificial Neural Networks in Pattern Recognition. In: Artificial Immune Systems: A Novel Paradigm to Pattern Recognition, pp. 67–84. University of Paisley Press (2000)
59. De Castro, L.N.: Dynamics of an artificial immune network. Journal of Experimental and Theoretical Artificial Intelligence 16(1), 146–153 (2004)
60. de Castro, L.N., Timmis, J.: An artificial immune network for multimodel funtion optimization. In: Proceedings of IEEE Congress on Evolutionary Computation, vol. 1, pp. 674–699 (2002)
61. de Castro, L.N., Timmis, J.: An Introduction to Artificial Immune System: A New Computational Intelligence Paradigm. Springer, Heidelberg (2002)
62. de Castro, L.N., Von Zuben, F.J.: An evolutionary immune network for data clustering. In: Proceedings of the IEEE Computer Society Press, SBRN 2000, vol. 1, pp. 84–89 (2000)
63. de Castro, L.N., Von Zuben, F.J.: Data Mining: A Heuristic Approach. In: aiNet: An Artificial Immune Network for Data Analysis, pp. 231–259. Idea Group Publishing, USA (2001)
64. De Castro, L.N., Von Zuben, F.J.: Immune and neural network models: Theoretical and empirical comparisons. International Journal of Computational Intelligence and Applications 1(3), 239–257 (2001)
65. de Castro, L.N., Von Zuben, F.J.: Learning and optimization using the clonal selection principle. IEEE Transaction on Evolutionary Computation, Special Issue on Artificial Immune Systems 6, 306–313 (2002)
66. Denny, M.: Introduction to importance sampling in rare-event simulations. Eur. J. Phys. 22, 403–411 (2001)
67. Dhillon, I.S., Guan, Y., Kulis, B.: Kernel k-means, spectral clustering and normalized cuts. In: Proc. ACM SIGKDD Intl Conf. Knowledge Discovery and Data Mining, Seattle, WA (2004), `http://www.cs.utexas.edu/users/inderjit/public_papers/kdd_spectral_kernelkmeans.pdf`
68. Diaconis, P., Freedman, D.: Asymptotics of graphical projections. The Annals of Statistics 12, 793–815 (1984)
69. Ding, C., Zhou, D., He, X., Zha, H.: R1-pca: Rotational invariant l1-norm principal component analysis for robust subspace factorization. In: Proceedings of the 23rd International Conference on Machine Learning, vol. 148, pp. 281–288 (2006)
70. Ding, H.Q., Ding, Z., He, X.F., Zha, H.Y.: R1-pca: rotational invariant l1-norm principal component analysis for robust subspace factorization. In: ICML, pp. 281–288. ACM, New York (2006)

71. Dunn, J.C.: A fuzzy relative of the isodata process and its use in detecting compact well-separated clusters. Journal of Cybernetics 3, 32–57 (1973)
72. Dybowski, R., Roberts, S.: Confidence Intervals and Prediction Intervals for Feed-Forward Neural Networks. Cambridge University Press, Cambridge (2001)
73. Dyrskjot, L.: Identifying distinct classes of bladder carcinoma using microarrays. Nat Genetics 33(1), 90–96 (2003)
74. Efron, B., Tibshirani, R.: An Introduction to The Bootstrap. Chapman and Hall, Boca Raton (1993)
75. Erwin, E., Obermayer, K., Schulten, K.: Self-organising maps: convergence properties and energy functions. Biological Cybernetics 67, 47–55 (1992)
76. Erwin, E., Obermayer, K., Schulten, K.: Self-organising maps: stationary states, metastability and convergene rate. Biological Cybernetics 67, 35–45 (1992)
77. Fan, C.Y., Wang, B.Q., Ju, H.: A new fastica algorithm with symmetric orthogonalization. In: 2006 International Conference on Communications, Circuits and Systems Proceedings, June 2006, vol. 3, pp. 2058–2061 (2006)
78. Fayyad, U., Bradley, P., Reina, C.: Scaling clustering algorithms to large databases. Technical Report MSR-TR-98-37, Microsoft, Redmon, WA (1998)
79. Fisher, R.A.: The use of multiple measurements in taxonomic problems. Annals of Eugenics 7, 179–188 (1936)
80. Fishman, G.S.: Monte Carlo: Concepts, Algorithms, and Applications. Springer, New York (1995)
81. Flexer, A.: Limitations of self-organizing maps for vector quantization and multidimensional scaling. In: Mozer, M.C., Jordan, M.I., Petsche, T. (eds.) Advances in Neural Information Procseeing Systems. Proceedings of the 1996 Conference, vol. 9, pp. 445–451. MIT Press, London (1997)
82. Flury, B., Riedwyl, H.: Multivariate Statistics: A practical approach. Cambridge University Press, Cambridge (1988)
83. Forrest, S., Perelson, A., Allen, L., Cherukuri, R.: Self-nonself discrimination in a computer. In: Proc. of the IEEE Symposium on Research in Security and Privacy, pp. 202–212 (1994)
84. Fraley, C., Raftery, E.: How many clusters? which clustering method?, answers via model-based cluster analysis. Technical Report 329, Dept. of Statistics. University of Washington (1998)
85. Friedman, J.H.: Exploratory projection pursuit. Journal of the American Statistical Association 82(397), 249–266 (1987)
86. Friedman, J.H.: Exploratory projection pursuit. Journal of the American Statistical Association 82(397), 249–266 (1987)
87. Friedman, J.H., Tukey, J.W.: A projection pursuit algorithm for exploratory daya analysis. IEEE Transactions on Computers c-23(9), 881–889 (1974)
88. Friess, T.-T., Cristianini, N., Campbell, C.: The kernel-adatron algorithm: a fast and simple learning procedure for support vector machines. In: Proceedings of the Fifteenth International Conference on Machine Learning, July 1998, pp. 188–196 (1998)
89. Fyfe, C.: Pca properties of interneurons. In: ICANN 1993, pp. 183–188 (1993)
90. Fyfe, C.: Introducing asymmetry into interneuron learning. Neural Computation 7, 1167–1181 (1995)
91. Fyfe, C.: A comparative study of two neural methods of exploratory projection pursuit. Neural Networks 10, 257–262 (1997)

92. Fyfe, C.: Hebbian Learning and Negative Feedback Networks. Springer, Heidelberg (2005)
93. Fyfe, C.: The topographic product of experts. In: Duch, W., Kacprzyk, J., Oja, E., Zadrożny, S. (eds.) ICANN 2005. LNCS, vol. 3696, pp. 397–402. Springer, Heidelberg (2005)
94. Fyfe, C.: Stochastic process methods for context-assisted information extraction. Technical report, University of Paisley (2006)
95. Fyfe, C.: Two topographic maps for data visualization. Data Mining and Knowledge Discovery 14, 207–224 (2007)
96. Fyfe, C.: Two topographic maps for data visualization. Data Mining and Kownledge Discovery 14, 207–224 (2007)
97. Fyfe, C., Baddeley, R.: Non-linear data structure extraction using simple hebbian networks. Biological Cybernetics 72, 533–541 (1995)
98. Fyfe, C., MacDonald, D., Lai, P.L., Rosipal, R., Charles, D.: Unsupervised Learning Using Radial Kernels, ch. 8, pp. 193–219. Springer, Heidelberg (2001)
99. Garcia-Osorio, C., Fyfe, C.: The combined use of self-organising maps and andrews curves. International Journal of Neural Systems 15(3), 197–206 (2005)
100. Garcia-Osorio, C., Fyfe, C.: Visualisation of high dimensional data via orthogonal curves. Journal of Universal Computer Science 11(4), 1806–1819 (2005)
101. Garrido, R., Rivero-Angeles, F.J., Martinez-Garcia, J.C., Martinez-Guerra, R., Gomez-Gonzalez, B.: Gradient algorithm for structural parameter estimation and nonlinear restoring forces. In: Proceedings of the American Control Conference, vol. 5, pp. 4770–4775 (2004)
102. Girolami, M.: Self-organising Neural-Networks for signal separation. PhD thesis, University of Paisley (1998)
103. Girolami, M.: Mercer kernel based clustering in feature space. IEEE Transactions on Neural Networks 13(3), 780–784 (2002)
104. Goldberg, D.E.: Genetic Algorithm in Search, Optimization and Machine Learning. Addison-Wesley, Reading (1989)
105. Goodman, D.E., Boggess, L., Watkins, A.: Artificial immune system classification of multiple-class problems. In: Proceedings of the Artificial Neural Networks in Engineering ANNIE (2002), pp. 179–183. Springer, Heidelberg (2002)
106. Grabmeier, J., Rudolph, A.: Techniques of cluster algorithms in data mining. Data Mining and Knowledge Discovery 6(4), 303–360 (2002)
107. Grefenstette, J.J., Ramsey, C.L., Schultz, A.C.: Learning sequential decision rules using simulation models and competition. Machine Learning 5, 355–381 (1990)
108. Han, J., Kamber, M.: Data Mining: Concepts and Techniques, 2nd edn. Elsevier, Amsterdam (2006)
109. Hartigan, J., Wang, M.: A k-means clustering algorithm. Applied Statistics 28, 100–108 (1979)
110. Hastie, T., Stuetzle, W.: Principal curves. Journal of the American Statistical Association 84, 502–516 (1989)
111. Hinton, G., Sejnowski, T.J. (eds.): Unsupervised Learning and Map Formation: Foundations of Neural Computation. MIT Press, Cambridge (1999)
112. Hinton, G.E.: To recognize shapes, first learn to generate images. Technical Report 004, University of Toronto (2006)
113. Hinton, G.E., Osindero, S., Teh, Y.: A fast learning algorithm for deep belief nets. Neural Computation 16, 1527–1554 (2006)

114. Hinton, G.E., Salakhutdinov, R.R.: Reducing the demensionality of data with neural networks. Science 313, 504–507 (2006)
115. Hoaglin, D.C.: Exploratory data analysis. Encyclopedia of Statistical Sciences 2, 579–583 (1982)
116. Holland, J.H.: Adaptation in Natural and Artificial Systems: An Introductory Analysis with Application to Biology and Control and Artificial Intelligence. University of Michigan Press (1975)
117. Hotelling, H.: Analysis of a complex of statistical variables into principal components. Journal of Educational Psychology 24, 417–441 (1933)
118. Hsieh, M.H.: Adaptive monte carlo methods for rare event simulations. In: Proceedings of the 2002 Winter Simulation Conference, vol. 1, pp. 108–115 (2002)
119. Huang, T.M., Kecman, V., Kopriva, I.: Kernel Based Algorithms for Mining Huge Data Sets, Supervised, Semi-supervised, and Unsupervised Learning. Springer, Heidelberg (2006)
120. Hyvarinen, A.: Noisy independent component analysis, maximum likelihood-estimation, and competitive learning. In: Neural Networks Proceedings, 1998. IEEE World Congress on Computational Intelligence, 1998, vol. 3, pp. 2282–2287 (1998)
121. Hyvarinen, A.: The fixed-point algorithm and maximum likelihood estimation for independent component analysis. Neural Processing Letters 10, 1–5 (1999)
122. Hyvarinen, A., Karhunen, J., Oja, E.: Independent Component Analysis. Wiley, Chichester (2001)
123. Jacobs, R.A., Jordan, M.I., Nowlan, S.J.: Adaptive mixures of local experts. Neural Computation 3, 79–87 (1991)
124. Jain, A.K., Dubes, R.C.: Algorithms for Clustering Data. Prentice Hall, Englewood Cliffs (1988)
125. Jardine, N., Sibson, R.: The construction of hierarchic and non-hierarchic classifications. The Computer Journal 11, 177–184 (1968)
126. Jerne, N.K.: Towards a network theory of the immune system. Annual Immunology 125C, 373–389 (1974)
127. Johnson, S.C.: Hierarchical clustering schemes. Psychometrika 2(3), 241–254 (1967)
128. Jolliffe, I.T.: Principal Component Analysis. Springer, Heidelberg (1986)
129. Jordan, M.I., Jacobs, R.A.: Hierarchical mixtures of experts and the em algorithm. Neural Computation 6, 181–214 (1994)
130. Kaelbling, L.P., Littman, M.L., Moore, A.W.: Reinforcement learning: A survey. Journal of Artifical Intelligence Research 4, 237–285 (1996)
131. Kannan, R., Vempala, S., Vetta, A.: On clusterings: good, bad, and spectral. Journal of the ACM (J. ACM) 51(3), 497–515 (2004)
132. Kaski, S., Kangas, J., Kohonen, T.: Bibliography of self-organizing map (som) papers: 1981-1997. Neural Computing Surveys 1, 102–350 (1998)
133. Kaski, S., Sinkkonen, J.: A topography-preserving latent variable model with learning metrics. In: Advances in Self-Organizing Maps, pp. 224–229 (2001)
134. Kaufman, L., Rousseeuw, P.J. (eds.): Finding Groups in Data: an Introduction to Cluster Analysis. John Wiley and Sons, Chichester (1990)
135. Keim, D.A., Kriegel, H.P.: Visualization techniques for mining large databases: A comparison. IEEE Transactions on Knowledge and Data Engineerring 8(6), 923–938 (1996)

136. Kepler, T.B., Perelson, A.S.: Somatic hypermutation in b cells: An optimal control treatment. Journal of Theoretical Biology 164, 37–64 (1993)
137. Kohonen, T.: Self-Organization and Associative Memory. Springer, Heidelberg (1984)
138. Kohonen, T.: Self-Organising Maps. Springer, Heidelberg (1995)
139. Koza, J.R.: Genetic Programming: on the Programming of Computers by Means of Natural Selection. MIT Press, Cambridge (1992)
140. Kroese, D.P., Rubinstein, R.Y.: The Cross Entropy: A Unified Approach to Combination Optimization, Monte-Calo Simulation and Mechine Learning. Springer, Heidelberg (2004)
141. Van Laerhoven, K.: Combining the self-organizing map and K-means clustering for on-line classification of sensor data. In: Dorffner, G., Bischof, H., Hornik, K. (eds.) ICANN 2001. LNCS, vol. 2130, pp. 464–469. Springer, Heidelberg (2001)
142. Lai, P.L.: Multiple canonical correlation with neural networks. Technical report, University of Paisley (2002)
143. Lai, P.L.: Neural Implementations of Canonical Correlation Analysis. PhD thesis, University of Paisley (2002)
144. Lai, P.L., Fyfe, C.: A neural network implementation of canonical correlation analysis. Neural networks 12, 1391–1397 (1999)
145. Lai, P.L., Fyfe, C.: Kernel and nonlinear canonical correlation analysis. International Journal of Neural Systems 10(5), 365–377 (2000)
146. Lawrence, N.: Probabilistic non-linear principal component analysis with gaussian process latent variable models. Journal of Machine Learning Research 6, 1783–1816 (2005)
147. Leen, G., Fyfe, C.: A gaussian process latent variable model formulation of canonical correlation analysis. In: Proceedings of the 14th European Symposium on Artificial Neural Networks, pp. 413–418 (2006)
148. Leinhardt, G., Leinhardt, S.: Exploratory data analysis: New tools for the analysis of empirical data. Review of Research in Education 8, 85–157 (1980)
149. Likas, A.: A reinforcement learning approach to on-line clustering. Neural Computation 11(8), 1915–1932 (1999)
150. Likas, A.: A reinforcement learning approach to on-line clustering. Neural Computation 11(8), 1915–1932 (2000)
151. Likas, A., Vlassis, N., Verbeek, J.J.: The global k-means clustering algorithm. Pattern Recognition 36, 451–461 (2003)
152. Lloyd, S.P.: Least squares quantization in pcm. Technical note, Bell Laboratories (1957); Published in 1982 in IEEE Transactions on Information Theory 28, 128–137 (1982)
153. Lovejoy, W.S.: A survey of algorithmic methods for partially observable markov decision processes. Annals of Operation Research 28, 47–66 (1991)
154. Luxburg, U.V.: A tutorial on spectral clustering. Technical Report TR-149, Max Planck Institute for Biological Cybernetics (March 2007)
155. Ma, X.L., Likharev, K.K.: Global reinforcement learning in neural networks with stochastic synapses. In: International Joint Conference on Neural Networks, 2006, July 2006, pp. 47–53 (2006)
156. MacKay, D.J.: Information Theory, Inference, and Learning Algorithms. Cambridge University Press, Cambridge (2003)
157. MacKay, D.J.C.: Maximum likelihood and covariant algorithms for independent component analysis. Technical report, Dept. of Physics, Cambridge University (December 1996)

158. MacQueen, J.: Some methods for classification and analysis of multivariate observations. In: Proceedings of the fifth Berkeley symposium on mathematical statistics and probability, pp. 281–297 (1967)
159. Mardia, K.V., Kent, J.T., Bibby, J.M.: Multivariate Analysis. Academic Press, London (1979)
160. Mataric, M.J.: Reward function for accelerated learning. In: Proceedings of the 11th International Conference on Machine Learning, pp. 181–189 (1994)
161. McGlinchey, S., Pena, M.: Quantization errors in the harmonic topographic mapping. In: The 9th WSEAS International Conference on applied mathematics, MATH 2006, May 2006, pp. 105–110 (2006)
162. McGlinchey, S., Pena, M., Fyfe, C.: Comparison of quantization errors in the model- and data-driven harmonic topographic mappings. WSEAS Transactions on Computers 5(7), 1562–1570 (2006)
163. Melzer, T., Reiter, M., Bischof, H.: Nonlinear feature extraction using generalized canonical correlation analysis. In: Proceedings of the International Conference on Arttificial Neural Networks, pp. 353–360. Springer, Heidelberg (2001)
164. Metropolis, N., Ulam, S.: The monte carlo method. Journal of the American Statistical Association 44(247), 335–341 (1949)
165. Michail, G.L., Parr, R.: Reinforcement learning as classification: leveraging modern classifiers. In: Proceedings of the 20th International Conference on Machine Learning, pp. 424–431 (2003)
166. Michie, D., Spiegelhalter, D.J., Taylor, C.C. (eds.): Machine learning, neural and statistical classification. Ellis Horwood (1994)
167. Mika, S., Ratsch, G., Weston, J., Scholkopf, B., Muller, K.-R.: Fisher discriminant analysis with kernels. Neural Networks for Signal Processing, 41–48 (1999)
168. Mika, S., Scholkopf, B., Knirsch, P., Muller, K.R., Smola, A.J., Ratsch, G.: Input space vs feasure space in kernel-based methods. IEEE Transactions on Neural Networks 10, 1000–1017 (1999)
169. Mika, S., Scholkopf, B., Smola, A., Muller, K.R., Ratsch, G.: Kernel pca pattern reconstruction via approximatate pre-images. In: Proceesings of 8th International Conference on Artificial Neural Networks, pp. 147–152. Springer, Heidelberg (1998)
170. Mika, S., Scholkopf, B., Smola, A., Muller, K.R., Scholz, M., Ratsch, G.: Kernel pca and de-noising in feature space. In: Advances in Neural Processing Systems, vol. 11, pp. 536–542 (1999)
171. Moreau, E., Macchi, O.: New self-adaptive algorithms for source separation based on contrast functions. IEEE Signal Proc., 215–219 (1993)
172. Moriarty, D.E., Schultz, A.C., Grefenstette, J.J.: Evolutionary algorithms for reinforcement learning. Journal of Artificial Intelligence Research 11, 241–276 (1999)
173. Muller, K.R., Mika, S.: An introduction to kernel-based learning algorithms. IEEE Trans. on Neural Networks 12(2), 181–202 (2001)
174. Ng, A.Y., Jordan, M., Weiss, Y.: On spectral clustering: Analysis and an algorithm. In: Proc. of NIPS-14, pp. 849–856 (2001), `http://ai.stanford.edu/~ang/papers/nips01-spectral.ps`
175. Oja, E.: A simplified neural model as a principal component analysis. Journal of Mathematical Biology 16, 267–273 (1982)

176. Oja, E.: Principal component and subspaces. International Journal of Neural Systems 1, 61–68 (1989)
177. Oja, E.: The nonlinear pca learing rule in independent component analysis. Neurocomputing 17, 25–46 (1997)
178. Oja, E., Karhunen, J.: Signal separation by nonlinear hebbian learning. In: Proceedings IEEE ICNN 1995, pp. 83–87 (1995)
179. Oja, M., Kaski, S., Kohonen, T.: Bibliography of self-organizing map (som) papers: 1998-2001 addendum. Neural Computing Surveys 3, 1–156 (2003)
180. Pampalk, E.: Limitations of the som and gtm (2001), `http://citeseer.ist.psu.edu/pampalk01limitations.html`
181. Pena, M., Barbakh, W., Fyfe, C.: Topology-Preserving Mappings for Data Visualisation. In: Principal Manifolds for Data Visualization and Dimension Reduction, pp. 132–152. Springer, Heidelberg (2007)
182. Pena, M., Fyfe, C.: Developments of the generalised harmonic topographic mapping. WSEAS Transactions on Computers 4(11), 1548–1555 (2005)
183. Pena, M., Fyfe, C.: Faster clustering of complex data with generalised harmonic topographic mapping (g-hatom). In: 5th WSEAS International Conference on Applied Informatics and Communications, WSEAS AIC 2005, pp. 270–275 (2005)
184. Pena, M., Fyfe, C.: The harmonic topographic map. Technical Report 35, School of Computing, University of Paisley (2005)
185. Pena, M., Fyfe, C.: The harmonic topographic map. In: The Irish conference on Artificial Intelligence and Cognitive Science, AICS 2005, pp. 245–254 (2005)
186. Pena, M., Fyfe, C.: Model- and data-driven harmonic topographic maps. WSEAS Transactions on Computers 4(9), 1033–1044 (2005)
187. Pena, M., Fyfe, C.: Tight clusters and smooth manifolds with the harmonic topographic map. In: 5th WSEAS International Conference on Simulation, Modelling and Optimization, WSEAS SMO 2005, pp. 508–513 (2005)
188. Pena, M., Fyfe, C.: Forecasting woth topology preserving maps: Harmonic topographic map and topographic product of experts application. In: First International Conference on Multidisciplinary Information Sciences and Technologies, InSciT 2006, pp. 42–46 (2006)
189. Pena, M., Fyfe, C.: Outlier identification with the harmonic topographic mapping. In: European Symposium on Artificial Neural Networks, ESANN 2006, April 2006, pp. 289–295 (2006)
190. Ritter, H., Kohonen, T.: Self-organising semantic maps. Biological Cybernetics 61, 241–254 (1989)
191. Romdhani, S., Gong, S., Psarrou, A.: A multi-view nonlinear active shape model using kernel pca. In: BMVC 1999, pp. 483–492 (1999)
192. Rosipal, R., Trejo, L.J.: Kernel partial least squares regression in reproducing kernel hilbert space. Journal of Machine Learning Research 2, 97–123 (2001)
193. Rubinstein, R.Y.: The simulated entropy method for combinaltorial and continous optimization. Methodology and Computing in Applied Probability 2, 127–190 (1999)
194. Rubinstein, R.Y.: Combinaltorial optimization, cross entropy, ants and rare events. In: Stochastic Optimization: Algorithms and Applications, pp. 304–358 (2001)
195. Rubinstein, R.Y.: Optimization of computer simulation models with rare events. European Journal of Operations Reasearch 99, 89–112 (1997)

196. Sallans, B., Hinton, G.E.: Reinforcement learning with factored states and actions. Journal of Machine Learning Research 5, 1063–1088 (2004)
197. Scholkopf, B., Smola, A., Muller, K.-R.: Nonlinear component analysis as a kernel eigenvalue problem. Neural Computation 10, 1299–1319 (1998)
198. Scholkopf, B., Smola, A., Muller, K.R.: Nonlinear component analysis as a kernel eigenvalue problem. Neural Computation 10(5), 1299–1319 (1998)
199. Scholkopf, B., Smola, A., Muller, K.R.: Kernel Principal Component Analysis. In: Advances in Kernel Mothods- Support Vector Learning, pp. 327–370. MIT Press, Cambridge (1999)
200. Seiden, P.E., Celada, F.: A model for simulating cognate recognition and response in the immune system. Journal of Theoretical Biology (158), 329–357 (1992)
201. Shawe-Taylor, J., Cristianini, N.: Kernel Methods for Pattern Analysis. Cambridge University Press, Cambridge (2004)
202. Shi, J., Malik, J.: Normalized cuts and image segmentation. IEEE Transactions on Pattern Analysis and Machine Intelligence 22(8), 888–905 (2000)
203. Shi, J., Malik, J.: Normalized cuts and image segmentation. IEEE Trans. Pattern Analysis and Machine Intelligence 22(8), 888–905 (2000)
204. Shihab, A.I.: Fuzzy clustering algorithms and their application to medical image analysis. PhD thesis, Department of Computing, University of London (December 2000)
205. Singh, S., Jaakkola, T., Littmann, M., Szepasvfiri, C.: Convergence results for single-step on-policy reinforcement learning algorithms. Machine Learning 38, 287–308 (2000)
206. Skurichina, M., Robert Duin, P.W.: Bagging, boosting and the random subspace method for linear classifiers. Pattern Analysis and Applications 5(2), 121–135 (2002)
207. Smola, A.J., Mangasarian, O.L., Scholkopf, B.: Sparse kernel feature analysis. Technical Report 04, Univeristy of Wiscosin (1999)
208. Stanley, K.O., Miikkulainen, R.: Efficient evolution of neural network topologies. Evolutionary Computation 2, 1757–1762 (2002)
209. Sutton, R.S.: Learning to predict by the methods of temporal differences. Machine Learning 3, 9–44 (1988)
210. Sutton, R.S.: Generlization in reinforcement learning: Successful examples using sparse coarse coding. In: Advances in Neural Information Processing Systems, vol. 8, pp. 1038–1044 (1996)
211. Sutton, R.S., Barto, A.G.: Reinforcement Learning: An Introduction. MIT Press, Cambridge (1998)
212. Suykens, J.A.K., Vandewalle, J.: Least squares support vector machine classifiers. Neural Processing Letters 9(3), 293–300 (1999)
213. Svensen, M.: GTM: The Generative Topographic Mapping. PhD thesis, Aston University, Birmingham, UK (1998)
214. Tan, P.-N., Steinbach, M., Kumar, V.: Introduction to Data Mining. Pearson Addison Wesley, London (2005)
215. Tang, N., Vemuri, V.R.: An artificial immune system approach to document clustering. In: Proceedings of the Twentieth ACM Symposium on Applied Computing, pp. 918–922. ACM, New York (2005)
216. Timmis, J.: Artificial Immune Systems: A Novel Data Analysis Technique Inspired by the Immune Network Theory. PhD thesis, Department of Computer Science, University of Wales (2000)

217. Timmis, J., Edmonds, C.: A comment on opt-aiNET: An immune network algorithm for optimisation. In: Deb, K., et al. (eds.) GECCO 2004. LNCS, vol. 3102, pp. 308–317. Springer, Heidelberg (2004)
218. Tino, P., Nabney, I.: Hierarchical gtm: constructing localized non-linear projection manifolds in a principled way (IEEE) Transaction on Pattern Analysis and Machine Intelligence 24(5), 639–656 (2001)
219. Tipping, M., Bishop, C.M.: Probabilistic principal component analysis. Journal of the Royal Statistical Society B 6(3), 611–622 (1999)
220. Tryon, R.C., Bailey, D.E.: Cluster Analysis. McGraw-Hill, New York (1973)
221. Tufte, E.R.: The visual Display of Quantitative Information. Graphics Press, Cheshire (1983)
222. Tukey, J.W.: Exploratory Data Analysis. Addison-Wesley, Reading (1977)
223. van Gestel, T., Suykens, J.A.K., Brabanter, J.D., Moor, B.D., Vandewalle, J.: Kernel canonical correlation analysis and least squares support vector machines. In: Proceedings of the International Conference on Artificial Neural Networks, pp. 381–386. Springer, Heidelberg (2001)
224. Villalobos-Arias, M., Coello Coello, C.A., Hernández-Lerma, O.: Convergence analysis of a multiobjective artificial immune system algorithm. In: Nicosia, G., Cutello, V., Bentley, P.J., Timmis, J. (eds.) ICARIS 2004. LNCS, vol. 3239, pp. 226–235. Springer, Heidelberg (2004)
225. Watkins, C.J.C.H.: Learning from Delayed Rewards. PhD thesis. Cambridge University, Cambridge (1989)
226. Watkins, C.J.C.H., Dayan, P.: Q-learning. Machine Learning (1992)
227. Wegman, E.J.: Geometric methods in statistics (2003), `http://www.galaxy.gmu.edu/stats/syllabi/csi877/Lecture1.pdf`
228. Whitley, D., Dominic, S., Das, R., Anderson, C.W.: Genetic reinforcement learning for neuroncontrol problems. Machine Learning 13, 259–284 (1993)
229. Whittle, P.: On principal components and least square methods of factor analysis. Skandinavisk Aktuarietidskrift, 223–239 (1952)
230. Williams, R.J.: Simple statistical gradient-following algorithms for connectionist reinforcement learning. Machine Learning 8, 229–256 (1992)
231. Williams, R.J., Pong, J.: Function optimization using connectionist reinforcement learning algorithms. Connection Science 3, 241–268 (1991)
232. Wu, Y., Fyfe, C.: Pre-processing using topographic mappings. In: ICNNB 2005, vol. 3, pp. 1881–1884 (2005)
233. Wu, Y., Fyfe, C.: Topology preserving mappings using cross entropy adaptation. In: International Conference on artificial intelligence, knowledge engineering and data bases (2008)
234. Wu, Y., Fyfe, C., Lai, P.L.: Stochastic weights reinforcement learning for exploratory data analysis. In: International Conference on Artificial Neural Networks, vol. 1, pp. 668–676 (2007)
235. Xu, L.: Least mean square error reconstruction principle for self-organizing neural-nets. Neural Networks 6, 627–648 (1993)
236. Yamauchi, B.M., Beer, R.D.: Sequential behavior and learning in evolved dynamical neural networks. Adaptive Behavior 2, 219–246 (1993)
237. Young, G.: Maximum likelihood estimation and factor analysis. Psychometrika 6, 49–53 (1940)
238. Yu, S.X., Shi, J.: Multiclass spectral clustering. In: Proc. of the 9th IEEE International Conference on Computer Vision, vol. 1, pp. 313–319 (2003)

239. Zaiane, O.R.: Data Clustering. In: Principles of Knowledge Discovery in Databases, ch. 8, http://www.cs.ualberta.ca/~zaiane/courses/cmput690/slides/Chapter8/index.html
240. Zhang, B.: Generalized k-harmonic means – boosting in unsupervised learning. Technical Report HPL-2000-137, HP Laboratories, Palo Alto (October 2000)
241. Zhang, B.: Generalised k-harmonic means- dynamic weighting of data in unsupervised learning. In: First SIAM International Conference on Data Mining (2001), http://www.siam.org/meetings/sdm01/pdf/sdm01_06.pdf
242. Zhang, B.: Comparison of the performance of center-based clustering algorithms. In: Advanced in Knowledge Discovery and Data Mining: 7th Pacific-Asia Conference, PAKDD 2003, pp. 63–74 (2003)
243. Zhang, B.: Regression clustering. In: Third IEEE International Conference on Data Mining (ICDM 2003), pp. 451–458 (2003)
244. Zhang, B., Hsu, M., Dayal, U.: K-harmonic means - a data clustering algorithm. Technical Report HPL-1999-124, HP Laboratories, Palo Alto (October 1999)
245. Zhang, K., Chan, L.W.: Dimension reduction as a deflation method in ica. SPLetters 13(1), 45–48 (1997)
246. Zhao, L., Liu, Z.M.: A genetic algorithm for reinforcement learning. In: Proceedings of the 1996 IEEE International Conference on Neural Networks, vol. 2, pp. 1056–1060 (1996)
247. Zhong, S., Khoshgoftaar, T., Seliya, N.: Clustering-based network intrusion detection. International Journal of Reliability, Quality and Safety Engineering (2005)

Index

MIX
Papier aus verantwortungsvollen Quellen
Paper from responsible sources
FSC® C105338

If you have any concerns about our products,
you can contact us on
ProductSafety@springernature.com

In case Publisher is established outside the EU,
the EU authorized representative is:
Springer Nature Customer Service Center GmbH
Europaplatz 3, 69115 Heidelberg, Germany

Printed by Libri Plureos GmbH
in Hamburg, Germany